# MANUFACTURING CONTROL

# MANUFACTURING CONTROL
## The Last Frontier for Profits

GEORGE W. PLOSSL

**George W. Plossl & Co., Inc.**
**Decatur, Georgia**

Reston Publishing Company, Inc./Reston, Virginia
A Prentice-Hall Company

**Library of Congress Cataloging in Publication Data**

Plossl, G   W
    Manufacturing control: the last frontier for profits.

    Includes bibliographical references.
    1. Production control. 2. Inventory control.
I. Title.
TS157.P53                          658.5'6                          73-8965
ISBN 0-87909-483-4

© 1973 by
Reston Publishing Company, Inc.
*A Prentice-Hall Company*
Reston, Virginia 22090

15 14 13 12

Printed in the United States of America.

*In any field, the professionals set the standards. Their understanding of principles, the language and the body of knowledge, their familiarity with techniques and their skill in doing something to make tomorrow better than today are the underlying grounds for all progress.*

*Working with professional managers in many companies developing better control of manufacturing has been my most rewarding and challenging experience. This book is dedicated to them. It contains some examples of their work—brief testimony to their accomplishments. Its purpose is to help them expand their efforts. It presents the story of how the pros work and what they are capable of doing. The potential benefits to individuals, their companies and our economic system from professionalism in this field are enormous.*

# Preface

If you were unfamiliar with the role of top management and with its principal problems, the public press would be of little help developing an understanding. It is concerned primarily with the social and political issues of the day—the economic problems of business get little attention until they become social and political issues. In the public press, top management is portrayed as high-salaried, mysterious, characterless individuals dedicated to preserving the status quo and negative to social and political progress. Management itself does little to overcome this image. They speak out on the issues of the day only under extreme provocation and, when they do, their utterances seem intended only for other management people, being couched in language hardly understandable to the general public. No one questions seriously that management needs social consciousness and political awareness, but first they need to insure the economic health of their operations and *this means profits.*

It is even difficult to get an appreciation for their real problems by looking at the technical and trade journals supposedly written for them. *Fortune, Harvard Business Review, Dun's, Business*

*Week* and similar management publications cater to
management's broad interests rather than to its real problems.
Studying the content of these journals you would think that
management was a group of dilettantes dabbling in a welter of
subjects from government regulations to human psychology, from
unstructured crystal-ball techniques to highly sophisticated
mathematical algorithms on high-powered computers and from
helpless reaction to a myriad of outside pressures to omnipotence
over all phases of corporate life. Peter Drucker's title "The Art of
Management" would seem to sum it up; there is apparently too
little science connected with it.

To those in management, however, there is little mystery about
what their real problems are. Giving good customer service, using
the least amount of capital in doing it and earning an adequate
profit on this investment are the common problems in developing
and maintaining the economic health of a business enterprise. In
spite of the considerable talk of social responsibilities and the need
for political involvement, our economic system still measures and
rewards or punishes a business based on its return on capital invested.

It should be equally obvious (although it is hardly ever discussed)
that business can play its social and political roles only when it is
economically healthy. Unless it is giving adequate wages to its
employees, paying satisfactory dividends to its owners and
providing acceptable service to its customers, a business will not be
around long to indulge in social and political activities. Meeting its
three objectives of good customer service, minimum capital investment and profitable operation must be the primary interest of
all managers.

As has already been said, most management literature deals with
the fringes and trivia related to these main problems. In the recent
past, however, there have been some very significant developments
in the field of inventory and production control resulting in signal
improvements in control of manufacturing in a few companies. The
basic principles in the field have been identified and some new
techniques developed which, in conjunction with older, proven
tools, make it possible for management to reach desired levels of
customer service, make dramatic reductions in inventories required
and also cut the cost of operations significantly. These are not
highly sophisticated, mathematical techniques but rather fundamental approaches available to small as well as large companies;
they require no special skills and are easy for the average business
manager to understand.

Since most of these approaches were developed by practitioners rather than college professors and other theorists who have time to publish, there has been a minimum of coverage in management literature. The purpose of this book is to present these concepts, to describe the techniques and to illustrate their application in business English. Its objective is to put into management's hands the tools they need to solve their real problems. These tools have been proven in practice. They will work for you as they have for others.

George W. Plossl

# Contents

# List of Illustrations

## TABLES

# 1

# What Is Manufacturing Control?

If your company could buy a "machine" which not only paid for itself in less than one year *but also reduced your net assets*, it would snap up this opportunity! Strange as it may sound, there is an "investment" which actually reduces capital required—it's called a "Manufacturing Control System" and its objectives are better management of inventories, tighter control of production costs and improved customer service.

There is *literally nothing* a management team can do that approaches the potential benefits of improved control of manufacturing and inventories. It slashes costs, compared to whittling away at them with product redesign, marketing promotions, methods improvements, new wage incentives, plant layout changes and the other familiar "profit improvement" techniques.

Here are some examples of savings achieved by companies who have improved control of their operations:

| Company | Annual Savings | Annual Sales |
|---------|---------------|--------------|
| A | $430,000 | $21 million |
| B | $770,000 | $32 million |
| C | $1,200,000 | $125 million |

Proper inventory control also slices investment in inventories. A company with about $25 million invested in inventories estimated that they would have *$9 million more* if they were still ordering materials the way they did prior to installing their improved manufacturing control program. An electromechanical equipment manufacturer with a successful control system concluded recently that his company now has *$7 million less* inventory than it would have if they still ordered and controlled materials in the old way.

Do you think this couldn't happen in your company? Think again! Dramatic benefits have been achieved in small, medium, and large companies—some of which thought they were well-controlled, too. Do you know of any other way you can attack both sides of the "return-on-investment" ratio simultaneously? How else can you improve profits and reduce investment at the same time? Not to mention giving better customer service too.

In spite of increased social pressures and political involvement, management's primary problems continue to be getting adequate capital, increasing productivity, and improving customer service. The manufacturing control system, dealing as it does with inventory management and production control, addresses all three of these objectives.

In one of the rare useful articles appearing in the *Harvard Business Review*, Dean Ammer reported a management consensus that "Cash may be in short supply in the next 10 years." His article, "What Businessmen Expect From The 70's", in the January-February 1971 issue also pointed out that about one third of all proposed capital investments are rejected for lack of sufficient funds. It is well known that lack of capital is limiting the growth of at least one company out of five, particularly the smaller, fast-growing ones we count on to develop new technologies and provide additional jobs for our growing work force. *Yet untold millions continue to be tied up in useless inventory!*

*Industry Week* in its October 4th, 1971 article, "Is There Still Time to Save U.S. Industry?" stated that "to become competitive with other producers in the world...our productivity must be increased to help offset our higher hourly employment costs which are now two to three times—and more—greater than those of other countries with which we compete." Extremely heavy capital investments in machine tools and equipment for improving productivity exceed the resources of many companies, and rapidly changing technology makes the economic life of much of this equipment disastrously short. We have thoroughly wrung out the benefits of individual incentive programs; in fact, many companies with mature programs find them more of a hindrance than a help in improving productivity. *Yet better planning and scheduling which make possible improved productivity without any fixed capital investment receive too little attention.*

A tragicomedy worthy of Shakespeare is the way many managers go about achieving these basic objectives of improved customer service, more profitable operation, and reduced capital investment. Using the time-tested technique called "management by objectives," these managers concentrate on one major objective at a time. First the objective is to "cut inventory investment." An arbitrary goal is set and the word goes out to everyone involved with inventories to take whatever actions are necessary to reach this target. Purchased material orders are cancelled or rescheduled, cuts are made in the direct labor force, manufacturing runs are shortened, excess materials are reworked and the production control department works overtime to pinpoint surplus materials and direct the actions necessary to cut the inventories. Does this concerted effort really work? Of course it does. The inventories come down somewhere near the desired goal, and management smiles its satisfaction and turns its attention to other problems.

Customer service isn't too good. The out-of-stock list is high and customers are complaining about missed delivery dates and too many repromises. Competition is making inroads with some old established customers and something must be done about it. Hence the objective now is "better customer service." People are hired to produce materials needed to fill back orders, overtime is used wherever necessary to fill shortages and meet delivery promises, expensive substitute materials and alternate operations are employed where necessary, and production control is still working overtime to identify shortages (and expedite them through the plant.) This program also works well. Customer service levels improve and the sales department is able to stop concentrating on complaints and do some constructive selling. Management turns to another objective.

Profit margins certainly aren't what they should be. Manufacturing costs, overtime, and indirect/direct ratios are obviously excessive. The new objective is "profit improvement." Departmental goals are established, teams are organized, the suggestion plan is overhauled and all efforts are concentrated on the new goal. "No more overtime," is the rule, indirect labor payrolls are slashed, machine set-ups are eliminated by combining and lengthening runs, cheaper substitute materials are sought, automation and innovation cut production costs and Sales helps out by agreeing to eliminate some low margin products from the line. Once again "management by objectives" is successful and the company can pay a nice dividend for the year.

This cycle is too common not to be familiar to most experienced managers. Its period varies from two to three years and is influenced by the business cycle and competition. Its amplitude is a function of top management's aggressiveness and the success of each stage. The lower the inventory goal and the shorter the time to achieve it, the greater the impact on customer service. The quicker service is restored and the higher its level, the greater the impact on costs and inventories. Slashing costs in a hurry inevitably increases inventory and harms customer service.

I call this the INSANE cycle. This acronym comes from the first letters of the words *IN*ventory, *S*ervice, *A*nd *N*et *E*arnings. It is truly insane in the sense of being unrealistic and unworkable in the real world. The three objectives of low inventory, high customer service and profitable operation are directly in conflict and cannot be attacked independently. Improvements cannot be made in one without compromises in the others. The reason is that inventory is the common denominator of all three: it's the largest manageable asset, it's necessary for customer service, and it's vital to economic operation. Sound management of inventories is the name of the game.

There is a growing awareness that inventory management

requires more than keeping a few simple records, using sound judgment, and applying a few well-known techniques like EOQ, statistical order points, and exponential smoothing—although some managers still believe it's that simple. A Sales Manager once said to me, "I think you experts are making this business too complicated. It seems to me the right approach is simple—just keep two months' supply of everything on the shelf. This way we'd have six turns, which is three times as good as we now have, our customer service would be right up where we want it if everything were on the shelf and shop problems should diminish if they planned to produce two months' supply."

The fallacies in this approach are not obvious to those unfamiliar with the principles of inventory planning, and it continues to beguile many managers. Unfortunately, such managers never think through the problems of *how to keep two months' supply* in stock— the products in trouble are usually the ones whose demand rate has changed significantly at a time when the company is unprepared for such changes. This approach ignores machine set-up and downtime required when two months' supply of everything is produced. It overlooks purchase discounts which are lost if larger quantities can't be ordered. It's an oversimplified solution to a complex problem.

On the other hand, many managers are overawed by the complexities of the problem. I talked to a Manufacturing Vice-President who had his people working hard to develop a mathematical model of their business. He saw logical relationships among forecast errors, inventories, and capacity. He really believed that what he needed to control manufacturing adequately was a series of mathematical equations relating enough variables to give him all the answers. He was living in no less a dream-world than the simplistic Sales Manager. No business in this dynamic world can be run by mathematical formulae making decisions for managers.

There are many good techniques in the field and some have been around for a very long time. The oldest is expediting, which started about 1880 when the first clerk was assigned to help a foreman get out production. All companies still do it, but it is literally amazing how few do any planning to do it right. In my years in the field, I have seen only one article written on how to plan for better expediting. No one thinks about it: we just pick people who know their way around the plant, have good feet, and a loud voice, and set them to work. As a result, expediting is wasteful of manpower and about as productive of real results as a can of worms.

The second oldest technique is machine loading. We've known how to do this since about 1900 when Frederick W. Taylor showed us how to set time standards and Henry Gantt popularized bar charts. Machine loading today is probably the most widely used useless technique in the field. Consuming thousands of tons of paper, machine load reports contribute scarcely a pound of benefit.

The third oldest technique is the Economic Order Quantity. EOQ has been around since 1915. My best estimate is that about eight times as many people get hurt from using EOQ's as ever identify specific savings from them. The field has been "Technique Happy" too long; it has concentrated on finding tools to do the job without a full understanding of what is necessary to make the tools work properly.

There are basic principles underlying successful application of the techniques and these can be violated only at your own peril. For a long time many managers thought that inventory management and production control required only good judgment and common sense. There is no area of management with a more diabolical tendency to turn common sense into deep trouble. Here are a few examples; see how many of these your company has used:

1/ *If a little expediting is good a little more will be better.* I know one company with 93 full-time expediters hard at work. This approach attempts to overpower the problems resulting from poor planning and control, and becomes self-defeating as soon as expediters begin to compete for capacity.

2/ *To get more production from a factory, you must put more orders into it.* Manufacturing people argue that you can't get out what you don't put in. I call this the "constipation cure"—if you're constipated, eat hearty, something has to give. Unfortunately, putting more work into a plant makes it even more difficult to get the right things out on time.

3/ *To get important jobs completed on time, get them started as soon as possible.* This is usually expressed as "give the plant plenty of time and they have no excuse for missing the date." Unfortunately, like releasing more orders, it just adds to work in process and makes it more difficult to get specific individual jobs out on time.

4/ *If the planned lead time isn't long enough, increase it.* This is really a logical method for getting into greater difficulty. You can't close the gap between planned and actual lead times by changing the planned figures; you must get more capacity, and work off the excesses causing the delay.

5/ *Lay out (stage) sets of parts further in advance of assembly if you need more time to expedite the shortages.* This will guarantee you will have more shortages to expedite. It reduces the flexibility of inventories, puts more work into the plant to compete with the true shortages and leads to "borrowing from Peter to pay Paul," which aggravates record errors, usually the basic problem which makes staging necessary.

6/ *If you're short several items made on one machine, get out of trouble quicker by cutting all the lot-sizes:* except that you'll really be getting into deeper trouble if your problem is lack of capacity in the machine to produce all you need of everything it makes.

These common sense approaches are only assaults on the symptoms. All really aggravate the basic disease. Like the driver who speeds up his car to get to the gas station before he runs out of fuel, the practitioner who applies "good common sense and sound judgment" without understanding the underlying principles in this field will soon find himself in deeper trouble.

What's necessary is that the techniques be used properly as part of a system. Although many, with some justification, look upon systems as an organized way to lose money, they are as necessary for good control of inventory and manufacturing as budgets are for expense control. As used here "system" means the structure of the planning and control activities and the formal records, procedures, and relationships which make this structure work effectively.

Better customer service, reduced costs, and lower inventory investment are conflicting objectives. My definition of "manufacturing control" is using an integrated system to:

1/ Help managers make realistic plans to achieve a desired balance among these three conflicting objectives.

2/ Measure actual progress in meeting these plans.

3/ Report significant deviations to those responsible for taking corrective action and,

4/ Act to correct the deviations or revise the plans.

This definition stresses the need for *realistic planning*. In *The Computerized Society* by James Martin and Adrian R.D. Norman (Prentice-Hall, 1970) the authors observed, "Our society has largely evolved by a process of chance mutation and adaptation, in the unplanned manner of natural selection. It reacts to major problems only when they starkly present themselves." Isn't this a fine description of the way many companies manage inventories and production? They move from crisis to calamity, always reacting to the latest stimulus.

Borg-Warner Corporation's York Division President Jack W. Kennedy, summed it up in his quotation in "The Mess in the Factory" (Dun's, June 1971): "Planning is still the name of the game. When you don't plan you're operating in an emergency situation with its inevitable high costs in off-schedule production, out-of-kilter inventories, overtime pay, out-of-wack deliveries and erratic supply." *True control is planning to stay out of trouble, not reacting to get out of it.*

The elements of an effective manufacturing control system have been identified only fairly recently. They will be discussed in detail in subsequent chapters, but briefly they are:

1/ Forecasting customer demand.

2/ Planning inventories:
    Classifying for degree of control.
    Deciding when and how much to order (planning priorities).
    Budgeting total inventories.

3/ Planning and controlling capacity (equipment and manpower).

4/ Controlling priorities:
    Scheduling and rescheduling.
    Work center loading.
    Dispatching and expediting.

To be effective, a system needs all of these elements. The impact of even minor deficiencies can be severe, just as a fine automobile can be wrecked because of defective windshield wipers. *The structure of this system and its elements are common to all businesses regardless of their form of organization, product, materials or processes.* The structure is influenced only to a slight degree by the way the product is manufactured. For example, make-to-stock businesses will place somewhat more emphasis on forecasting, and their system will tolerate longer delays in measuring what is actually going on in the factory compared to make-to-order companies.

The objective, of course, is not to have a fine system but to get control. This requires:

1/ Realistic goals, including demand forecasts, production schedules, inventory levels, plant capacity, and customer service levels.

2/ Practical tolerances on these goals so that the system can highlight by exception those activities not under control.

3/ Timely and accurate measurements of significant deviations from the plan (those activities outside the tolerance limits).

4/ Communicating these deviations to those responsible for taking corrective action.

5/ Taking effective action promptly.

There is nothing automatic about getting control with a system. It's a man/machine interaction in which men, the managers, must set the basic objectives, establish the tolerance limits, analyze the feedback from the system and insure that corrective action is taken.

A manufacturing control system is an *information system,* handling data such as forecasts, part numbers, bills of material, order quantities and scheduled dates. It is not "Materials Management"; its focus is on information, not materials. Recognizing this distinction represents a dramatic breakthrough. It puts the role of the computer into perspective, as will be discussed in Chapter 2. It suggests the significantly different approaches to organization which will be covered in Chapter 3. It provides a very different basis for measuring professional performance (the subject of Chapter 9).

Manufacturing control requires professional managers. It's a new game whose rules have only recently been defined, whose paraphernalia includes the computer and other devices strange to the touch, and whose tricks are still as unfamiliar to most of the players as they are to the spectators. The cost of making mistakes is becoming prohibitive but the potential benefits from a truly professional performance are tremendous. Some companies have already learned how to do the job effectively. If your company's control system is not effective you cannot afford to wait. A synonym for "business failure" is "a successful manufacturing control system in the hands of a competitor."

A company which has developed professional performance in this area has its competitors on the run. Regardless of the type of business, the kind of product, the materials used, or the processes employed there is nothing with the potential for improving profits and reducing investment that compares with better control of manufacturing and inventories. The potential returns are of orders of magnitude greater than new marketing promotions, product design improvements, better methods and standards, and the other common activities of profit improvement programs. The company that succeeds in this area achieves tremendous benefits; the company that does not may not survive.

# 2

# Laying
# the Foundations

The elements necessary for an effective manufacturing control system were presented in Chapter 1. They must be linked together in a basic structure including the records, decision rules and procedures which allow them to work together effectively. The sequence of activities to be followed in using the system to plan and control manufacturing is also independent of the nature of the business involved. How it must be done is shown in Figure 1. Here's how it works:

### Develop a Master Production Schedule

This is the master planning which relates the sales forecast, the financial plan, and the manufacturing plan so that all major departments are using the same strategy. At this step the basic information required for developing inventory budgets, direct labor totals and sales income is determined. This is a familiar procedure in chemical, pharmaceutical, basic metals and a few similar industries where capital investment is high per hour of

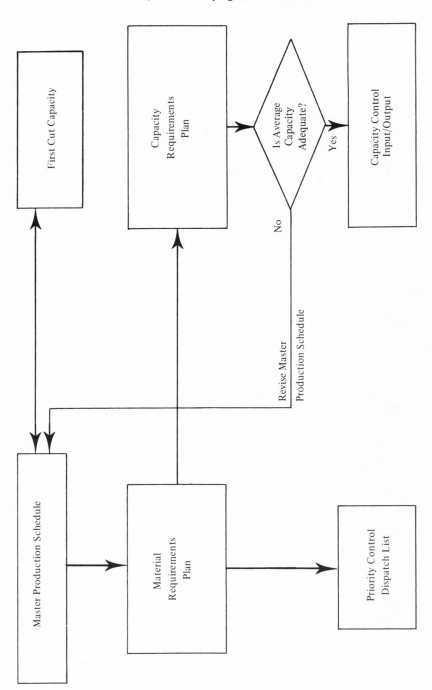

**Figure 1** / Sequence of Activities of Manufacturing Control.

direct labor. Few metal, plastics and woodworking manufacturers, however, recognize the need for a master production schedule. This is not a once-a-year activity; it must be done regularly, usually at monthly intervals but more frequently if the business is subject to rapid changes in demand.

*The master production schedule must recognize* capacity limitations which exist in short-range periods during which capacity cannot be changed significantly. It is the key to sound planning and control. It must state only what the plant *is and will be* capable of making—not what you'd like it to make.

## Plan Material Requirements

Scarce resources of men, time and money prevent exercising the same degree of control on all items in inventory. Attention must be focused on the "Vital Few" to achieve maximum return on the investment. Classifying the inventory identifies the Vital Few and, conversely, the Trivial Many which can be controlled by simple, inexpensive (but not undisciplined) methods. Two basic ordering systems are available to determine when and how much material should be ordered—the Order Point/Order Quantity System and Material Requirements Planning. A basic principle has been identified which determines where each should be used. *This step really determines the priorities of work, and ordering systems must be updated frequently, usually weekly, to keep these priorities valid.* Techniques (to be discussed in detail in Chapter 5) are now available which provide means for professional practitioners to develop trade-off information for management to aid them in deciding whether they want more or less inventory. These relate payback, intangible benefits or additional costs associated with changing the investment.

## Plan Capacity Requirements

This is probably the most vital of all activities since it determines the equipment and manpower necessary to meet projected customer demands and to achieve the desired changes in inventory totals. *It is the missing link in most manufacturing control systems.* Information, in the form of planned orders to be released in the future, coming out of the material requirements planning activities, provides part, but only part, of the data

needed to determine capacity requirements. Unplanned
requirements must also be met. These result from scrap and
rework, tooling problems, machinery or equipment breakdowns,
cost reduction changes, prototype production and a myriad of
other upsets. High precision is not essential; excellent results
have been achieved using fairly crude averages, "typical"
products and similar short-cuts.

## Capacity Control—Input Control

Only when work is released *at a rate and in a mix* which matches
the capacity of the plant can lead times be short and dependable.
These are essential since the ordering system depends on reliable
lead times to provide valid information about the dates material
is required. Contrary to popular opinion, *lead times are not
"ordained;" they can be controlled. Intelligent release of work*
even when capacity is inadequate is vital to such control.

## Capacity Control—Output Control

Getting the right total capacity at the right time is the most
essential requirement for achieving the desired levels of customer
service, plant operating costs and inventory investment. If your
plant (or your vendors) is not making enough in total, proper
priorities cannot be held on the specific items you need day-by-
day. If the desired capacity levels cannot be achieved, even for
the short range, it is essential that the master production
schedule be adjusted if priorities on individual replenishment
orders are to be at least realistic. This is no place for wishful
thinking. Successful companies plan for level production but
they are intolerant of delays in adjusting capacity when
necessary.

## Priority Control

These activities are needed to ensure that available capacity is
not wasted working on the wrong items. Dynamic priority
techniques are available to update current priorities daily.
However sound the planning, it alone will never be enough.
Murphy's Law, "What can go wrong will go wrong," was in-
vented in a factory. Manufacturing men are only too familiar

with O'Leary's corollary, "Murphy was an optimist." Customers will always reserve the right to change their minds and plants will always suffer from interruptions due to machine and equipment outages, absenteeism, tooling difficulties, process upsets, engineering problems and the like. The system must cope with these upsets and indicate promptly where corrective actions are needed or where plans must be revised.

This sequence of activities provides a closed-loop procedure for translating the Master Production Schedule into detailed plans, and then adjusting operations (or the plans) to achieve the output specified in the master schedule. The many individual techniques have been known and used for a long time. Only recently have we learned how to build these techniques into a system structure where they can work together. We've had the horses; we had to develop the harness to get them to pull together. The new ingredients are the recognition that a system structure is needed in which these techniques can function properly together, and data processing power to handle the masses of facts and figures.

The combination of masses of data and the need for rapid response to frequent changes rules out manual systems in practically all companies. The computer is a must, but it has not yet been an unmixed blessing to manufacturing people. We need to re-examine its role in manufacturing control.

## THE ROLE OF THE COMPUTER

Since its entry into the business world in the mid-1950's, the computer has been the tool of the scorekeepers, not the players. Controllers and treasurers recognized early the tremendous value of this machine in storing, manipulating and recording data at infinitesimal fractions of the cost of human labor and at electronic speeds. In spite of the heavy expense and the time-consuming systems design and programming effort required to put computers to work, financial people saw significant net gains. Their foresight has been responsible for practically all nonscientific, business computer installations. Inevitably they control them.

Line manufacturing, inventory management and production

control people, on the other hand, have been about as eager to get involved with computers as with dentists. If we depended on them to justify and use computers, we'd be running less than one percent of the number now in operation. *This is really most unfortunate since the real payback from using computers is in better control of inventories and production.*

Two reasons probably account for the reluctance of manufacturing people to use computers. For one, those who control computers and systems design have not understood manufacturing control problems. Because of this, computer applications they initiated, were either too simple or too sophisticated to be effective. Equally or even more important, production and inventory control people failed to understand the need for a control system—an information system. Preoccupied with materials, machines, and crises, most inventory and production control people saw little or no logic in their decision-making and consequently small application for the computer.

Now that the manufacturing control system has been defined, however, and the real function of inventory and production control people has been identified as information management, the role of the computer is clear. It is as unrealistic to contemplate good control of manufacturing without the computer as it is to think of good transportation without the airplane. There are still people who will not fly and there are many managers who want no part of the computer. This must change. Both the computer and the airplane are here to stay and failure to use them effectively can be a severe handicap. Sticking to manual systems can be fatal to a manufacturing operation.

The computer's fantastic ability to store data, manipulate it, and produce information for decision making is completely complementary to man's ability to analyze such information, draw proper conclusions and react quickly to change. The computer and the man make an unbeatable team; where one is strong the other is weak. This is illustrated in Figure 2. Because of its vital role in manufacturing control, control of the computer itself becomes an important question. Although he didn't intend to, Bernard Levin, columnist in the London *Daily Mail,* summed up the reaction of most managers to the computers in their companies. "I . . . believe that the computer is not the great god we have been lead to believe, but a hollow idol, manipulated by crafty priests." There is more

| Computer | Man | |
|---|---|---|
| Fast | Slow | in calculation |
| Accurate | Inaccurate | handling data |
| Dependable | Erratic | with routines |
| Inexpensive | Expensive | per transaction |
| Inflexible | Flexible | for change |
| Rigid | Adaptable | handling upsets |
| High | Low | total cost |
| Witless | Intelligent | in operation |

**Figure 2** / Characteristics—Computer vs. Man.

than a small resemblance between the men who feel they are the real producers for their company watching computer specialists with their mystical machines and the ancient tribal warriors watching their medicine men go through their mumbo-jumbo rituals. It is becoming more apparent every day that the large central computer and systems design staff is actually run by specialists and is incapable of responding to the real needs of manufacturing people. The blame for this certainly lies as heavily on manufacturing people as on the computer specialists.

Two solutions are possible. One is that manufacturing control people can get their own computer complete with systems design staff. This is not as farfetched as it may seem to those unfamiliar with computer hardware and software. The development of low-cost, mini-computers and packaged programs has not only reduced the investment required, it has also minimized the staff needed for systems and programming work. In my opinion, we will see a rapidly increasing number of companies, particularly small ones or small divisions of large corporations, following this route.

The other solution is to combine manufacturing control, computer control and cost accounting activities, recognizing that *they really are doing the same thing—providing information for control.* It really doesn't matter whether the information is expressed in pieces or dollars. It is easy to see how some other major problems could be solved by locking the cost system tightly to the inventory system. Who knows, this might even eliminate the annual write-off! This will be discussed further in Chapters 3 and 8.

## RECORD ACCURACY

If the job of managing inventories and controlling production is essentially managing information, then it is high time we paid some real attention to record accuracy. The old saw about the weather, "Everybody talks about it but nobody does anything," is certainly appropriate to record accuracy. Inventory records, bills of material, manufacturing routings and standards, open order files and similar records in most companies are about as reliable as an automobile clock. The true quantity on hand, the way the product is really put together, and the way parts are actually made are recorded in subsystem side-records and in the minds of people. Just putting files of bad data on the computer not only doesn't help, it frequently aggravates the accuracy problem because it introduces more chances for errors through data lost during transmission, key-punching mistakes and the like.

I have often wondered why it is that so many good managers who know their records are wrong do nothing about it. *Wouldn't it be simple if we looked at record accuracy, the foundation of our information system, the same way we look at building foundations?* Nobody questions the dirty, time-consuming, expensive work of building footings and foundation walls under his plant. In itself, the foundation obviously earns nothing and generates not one cent of return on the high capital investment in it. Managers react caustically to this comment, however. They all recognize clearly that, without a good foundation, the plant might fall down around their ears or they'd spend a lot of money propping it up to prevent its falling down. It would be a milestone if they recognized that record accuracy has the same function as a building foundation. Without it, the formal system falls down around your ears, or you spend valuable time and money propping it up with subsystems like assembly hot lists, foremen's black books, staging areas full of material for assembly, and similar shoring needed because the formal system can't cope.

After considerable investigation and thought I have concluded that there are two misunderstandings preventing the concentrated attack on record errors that most companies need. One is that the job of getting accurate records is a massive, expensive task; the other is that the payback from having accurate records just doesn't justify the effort. Both of these couldn't be more wrong. *The cost of having accurate records is only a small fraction of what most managers think.* The essential activities required are shown in Figure 3.

1. Establish a climate of high expectations.
2. Assign responsibility for counting, identification, auditing.
3. Provide adequate "tools", physical and system.
4. Measure record accuracy.
5. Find the causes of errors.

**Figure 3** / Steps to Get Accurate Records.

It is interesting to note that "higher wages" is not one of the factors. Many managers have the impression that they don't pay stockroom clerks, shop planners, expediters and other clerical people enough money to expect them to be accurate. O.A. Ohmann in his classic article "Skyhooks," which appeared twice in the *Harvard Business Review,* May-June 1955 and January-February 1970, hit the fallacy of this thinking right on the nose. He said, "Raising the price of prostitution does not make it the equivalent of love". Higher wages alone will never generate improved accuracy. Look at the bank clerk, probably the most precisely accurate individual we have, yet *paid less than most unskilled workers* in industry. He's accurate because of the way his job is organized. With him in mind, let's review the activities listed in Figure 3:

1/*Establish a climate of high expectations*—as long as management is tolerant of mistakes, people will continue to make them. *A zero defects goal is absolutely necessary.* Precision in a bank is a result of the climate of accuracy in which their people operate. To attain accurate records in business requires that management establish a similar climate: a complete intolerance of mistakes. What does this cost? Some management time to communicate the policy, some disciplinary action for those who fail to meet it, but *nothing out-of-pocket.*

2/*Assign responsibility*—in industry, plant people look on record errors as "an office problem" and the clerical force blames the workers in the plant for "having no real understanding of the vital need for accurate information." Both can prove that they are right but this does not solve the problem. *People who put data into the system must assume responsibility for its accuracy.*

Receiving clerks should be held accountable for identification of material received from vendors. You can't hold them responsible for the quantity unless you give them the time and the tools necessary to check the vendor's counts, if only on a sampling basis. If you excuse the receiving clerks, then you'll have to assign this responsibility to the storekeepers who put the material away or to the individuals using the materials.

Storekeepers should be responsible for quantities of materials received and issued, and for the location of all items in their stockroom. They should also be responsible for the identity of materials issued but not for identifying items received. The best place to hang this responsibility is on the production man who made the parts and was paid for making them. He should certainly know what he made.

Product design engineering should be responsible for the accuracy of bills of material describing to others how the product should be put together. Manufacturing engineering should be held accountable for routings and standards.

How much does this cost? Again *nothing out-of-pocket*—just some management time to identify where specific responsibility for record accuracy lies, and some discipline to enforce it.

3/ *Provide adequate tools*—bank clerks have desk calculators, change-making machines, an orderly storage space, and well-defined documents for each transaction, including a control check to determine whether they are right or wrong at the end of the day. How many of these kinds of tools do you find in the hands of people in industry handling transactions which affect the records? The most important single place affecting record accuracy, the stock room, resembles a "happening" in most companies. One stock room I looked at had material stored in every conceivable kind of container: a multitude of small cardboard cartons, fibre drums, wire-bound crates, wood boxes, large steel shop boxes—you name it, they had it.

A stockroom clerk's work station is usually an old discarded desk with a dull pencil and some battered boxes for his records. How many have you seen with machines or tools to help him count? Counting scales are the exception rather than the rule.

When they are used, the container tare weights, which really fix the accuracy of the count, are anybody's guess. Everyone has access to the storeroom but no one has responsibility to tell the clerk what he took. Significant quantities of material move directly from receiving or manufacturing departments to assembly areas without his even seeing them, and he sits with "uncashed" requisitions, wondering when he is going to get the material to fill them.

Here you could spend some money. I know of one company which spent about $22,000 to reorganize, enclose and lock up its storerooms. Replacing haphazard shop containers could cost several thousand dollars. You will probably get this back in fewer injuries, less rework of damaged parts and less maintenance on the containers themselves. Counting scales could cost a few hundred dollars. *This is really the only one of the activities required for accurate records on which you can spend a fair amount of money.* The total needed to do a good job, however, is guaranteed to be surprisingly small.

4/ *Measure Record Accuracy*—it is possible that the *principal reason management is so little concerned about record errors is that they have no specific measures of how sick they really are.* Many with whom I have talked assume that because the auditors accept the results of the annual physical inventory, the records must be in pretty good shape. Nothing could be further from the truth. The auditors merely certify that the *total is within reasonable limits of accuracy.* This says absolutely nothing about the individual item records. "Plus errors" and "minus errors" tend to offset each other so that the total can be reasonably close, while the individual items could be off by minus 100%/plus 300%.

Many managers also assume that the annual physical is adequate to correct the records and that is equally wrong. The way most companies take a physical inventory, the records are less accurate after the results have been posted than they were before. Counting errors affecting one record each are a significant problem with everyone rushing to get the inventory completed, but even more important are identification errors. They make *two records wrong.* In my opinion, the annual physical inventory should be outlawed as a means of adjusting individual item records, and discounted as a means of verifying that the assets of the company are accurately reflected in its records.

*Cycle counting, checking a sample of a few items at a time every day, every week, every month, is the only effective way to measure the accuracy of inventory records and to correct them.* The frequency of checking individual items can be associated with the frequency of activity of the item; an item of frequent transactions means more chances for error and requires more frequent checking. A few experienced people specializing in this type of physical inventory become familiar with locations and paperwork procedures and can check many items accurately. One man can easily check between 75 and 100 items per day; under ideal conditions a daily rate over 150 has been achieved. Cycle counting spreads the effort of taking physical inventories over the year, corrects records at frequent intervals, and establishes quantitative measures of record accuracy indicating whether or not improvement is being achieved. How much does this cost? Allowing even $12,000 annually for each person involved, it is obvious that cycle counting can be effective at a fraction of the cost of the annual physical inventory.

5/ *Find basic causes of errors—real improvement in record accuracy will only be made when the basic causes of errors are identified and eliminated.* Diagnosing such causes is no job for clerks; it requires the analytical ability and the time of people who have broad understanding of the records and systems and possible causes of failure to use them properly. The real diseases are usually found to be "error-prone" (like accident-prone) people, failure of people to follow the correct procedures, people taking short cuts and similar "people problems." *Systems do not make errors; people do.* Minimizing and eliminating them is a management problem, not a systems problem. How much does this cost? Out-of-pocket expense is again zero; the only cost is the time of those managers involved.

## WHERE'S THE PAYBACK FROM ACCURACY?

How does accuracy pay off? It's not really too difficult to assign some dollars-and-cents payback to reducing record errors. Figure 4 summarizes where some of the savings can be found.

1. No annual physical inventory.

2. No staging of sets of parts, materials.

3. No manual checking of counts.

4. Less obsolete inventory.

5. Reduced expenses from upsets.

**Figure 4** / Savings from Accurate Records.

1/ Probably the largest single saving is the elimination of the annual physical inventory, which can cost from $40,000 to $120,000 or more. *Companies with a high level of record accuracy no longer find it necessary to take an annual physical inventory;* in fact, they have concluded that such a mass effort with extreme pressures to get back into production cannot produce reliable inventory data, no matter what the cost.

2/ Staging the components, collecting them together physically in the storeroom prior to assembly, is the only way many companies have of knowing which parts are really missing. *With accurate records, staging can be done on paper,* and it is unnecessary to incur the extra handling charges, assign the extra space, tie up the inventory, and suffer the delays, cannibalizing, and other problems such staged materials generate. Based on the experiences of companies I have known, staging can cost from $10,000 to $30,000 per year.

3/ When records are accurate it is unnecessary to go to the plant and physically check materials before making important decisions like whether you can meet a special customer's delivery requirement, when is the best time to phase in a design change, or whether you should now shut down a machine or process for overhaul. This double checking can involve from $1,000 to $10,000 annually.

4/ When records are wrong, excess inventory often appears unexpectedly, long after a design change is made or a product is discontinued. It's too late then to use up these materials and the result is more obsolete inventory. From checks I have made it appears that 10 to 25% of obsolete inventories could be the result of poor records. The balance are due to poor inventory control, lack of control of the timing of engineering changes or market surprises.

5/ Accurate records also minimize the surprise excesses and unexpected shortages which make overtime and extra setups necessary. When plans and schedules are made on firm bases, assembly lines, machining operations and process equipment lose less time because of unexpected material shortages. Labor productivity, particularly in assembly areas, is increased markedly. The sum of such savings can easily exceed 3% of the direct labor cost.

While these tangible savings are quite significant, the intangible benefits can be even more so. The secret of management success is making sound decisions and this can only be done on the basis of accurate data. The year-end inventory write-off, so prevalent in most companies, is a symptom of failure to manage records and systems. These write-offs, rarely less than a quarter of a million dollars (one reached the astronomical level of fifteen million dollars), represent the cumulative effects of many transaction errors which cause the cost records to get "out of step" with inventory records. The system used for cost analysis and financial purposes uses different sources of data from the inventory record system. For example, inventory records for purchased materials are updated from receiving reports, while the dollar accounts for the same materials are adjusted by the vendors' invoices. *Designing the two systems to use the same input data and transactions makes it impossible for them to get out of step.* The data can be expressed in the units of measure (pieces or dollars) most meaningful to the managers needing it. What does it cost a company when profits are overstated, when management fails to recognize a problem affecting profits early in the game, or when valuable management time is taken at a critical period to find out why the write-off occurred? It may not be possible to assign dollar answers to these questions, but no manager would doubt that eliminating such problems was worthwhile.

Out-of-pocket expense is really insignificant, and the tangible savings alone represent a very handsome return on the effort needed to improve record accuracy. Corrective action to progress to better records should be a part of every management plan. I know of only four companies who can truthfully say that their records are accurate and prove it. This is a ridiculous situation. The basic question which management must answer is not "How much money can we save by making our records more accurate?" It is "Do we want to operate our business with an organized, formal system or with a collection of informal subsystems?" Do you depend on assemblers to know how the product is really put together? Do you use "hot lists" to identify the real priorities, and find the true shortages by physically segregating materials in advance of assembly? How sick you are is easily measured by how much you depend on such sub-systems to run your operation. *The alternative to accurate records is poor decisions.* Which can you afford?

# 3

# Organizing
# for Control

Since its inception, manufacturing control has been handicapped
by organizational problems. In the beginning it was under-
organized, its activities fragmented among manufacturing, industrial engineering, purchasing, and other parts of an
organization. This is still a major problem in European companies,
undoubtedly because of failure to recognize its primary funtion—
managing information. What this really means is making realistic
plans, getting timely and accurate measures of actual performance,
comparing them to the plan, and reporting significant deviations
promptly to those responsible for corrective action.

Many companies still suffer from having separate inventory
planning and production control functions. This split resulted from
the way the field developed. "Production Control" began when
clerks were added to assist foremen, who at that time hired and laid
off workers, ordered materials and set production schedules, thus
performing many inventory and production control functions. The
clerk kept his records on materials and manpower, contacted the
Sales Department and others interested in deliveries, and filled the

vital role of stockchaser through the plant. As his activities expanded into several manufacturing departments, the clerk was "promoted"; he then reported to a general foreman or superintendent. Later, as Production Control expanded further, working with inventory control, manufacturing engineering, tool cribs, receiving and other departments, it grew in importance in the organization. Where it exists today as a separate department, it usually reports to the plant manager or manager of manufacturing.

Inventory Control developed considerably later. People who understood mathematical analysis, probability, and statistical theory saw the application of these techniques to the problems of determining how much and when to order materials. The data and records necessary to apply these techniques were then transferred from the factory to an Inventory Control group in the office. Inventory Control's basic functions were keeping the necessary records and writing orders for materials. Their job, they believed, was finished when they released orders stating the necessary quantities and the dates wanted to vendors or to the factory. It was then someone else's job, usually Production Control's, to get these orders completed on time. Unfortunately, it never worked very well. Inventory and Production Control people both failed to understand how lead time is really controlled. Inventory *planning* cannot be effective unless it considers plant capacity and releases orders at a rate the plant is capable of handling.

Today, however, the pendulum has swung to the other extreme, and there have been major attempts to solve manufacturing control problems through reorganization. Inventory Control and Production Control were combined into a function frequently called Materials Control in the hopes that this would get materials delivered on time. This idea seemed to have so much merit that it was expanded to include all functions related to materials. Called "Materials Management," this brought together in one massive organizational group inventory and production control, purchasing, receiving, shipping, storekeeping, trucking, traffic, and in many cases, distribution and warehouse control. Its stated objectives were to provide one individual who could answer all questions relating to deliveries of materials, and achieve the maximum coordination among the functions affecting materials.

All it really did was add another organizational level and another high-priced manager to the payroll. It didn't solve the problems; it just buried them deeper in the organization. It never delivered a fraction of the benefits claimed for it. This is no surprise since it didn't address the real problems. Most Materials Managers were not professionals in the field, did not have adequate supporting systems, and didn't clearly understand their true role. They thought their problems were poor communication and poor coordination of material flow. Their problems were really poor planning and control of information. Materials Management brought one major benefit. It combined related activities in an organization structure within which sound systems can be developed. Ideas die hard, however, and many companies are still trying to make Materials Management work without the essentials needed to make it effective.

In his great little book, "Up The Organization," Bob Townsend discussed reorganizing, and quoted a Greek, Petronius Arbitor, who lived about 60 A.D. He said, "I was to learn later in life that we tend to meet any new situation by reorganizing; and a wonderful method it can be for creating the illusion of progress while producing confusion, inefficiency and demoralization." He couldn't be more right about what's happening today in many companies. *Reorganization is no substitute for a clear definition of primary functions and professional leadership with adequate systems tools to carry out these functions.*

To get people to work together effectively, there is no question that there must be some form of organization. This is necessary to achieve direction, motivation, discipline, administration, coordination, communication and all the other interactions needed if people are to work together toward some common goal. I don't propose here to add to the volume of literature already available on the theory and psychological implications of organization. Theory X/Theory Y, Kepner-Tregoe and other philosophies undoubtedly make some contribution. In my opinion, however, these concepts may be as necessary as sheet metal on an automobile but you need a sound chassis and power train underneath if you really want mileage.

The specific form of the organization doesn't appear to be really important. For proof just look at the great variety of different forms in use today among successful companies. In December 1970, Dun's published an analysis of 10 well-managed companies. One of them was ITT, and its success was attributed to "three Geneen laws—problems must be highlighted, facts must be unshakeable and communication must be face-to-face." The article further stated that ITT also operates with "a business plan, continual reporting of actual results as compared to plan and a procedure for remedial action when the two diverge." Isn't this exactly what I said in Chapter II about the way a control system works? Those familiar with ITT's great diversity of operations, including manufacturing transmission equipment, pumps, telephone switch gear, bakeries, car rental, and hotel operations know that its forms of organization are as different as the businesses. Why are they so successful? "Many others get the same numbers," says Harold Geneen, President, "but maybe they don't work as close to them or as quick at them." *Success requires an effective planning and control system with competent managers taking prompt action, not a specific form of organization.*

I think the really important factor is defining primary functions for the organization groups and then identifying areas of mutual responsibilities where joint efforts are necessary. Organizations are successful when they establish realistic objectives and get their people to pull together to solve mutual problems. Much has been written in the past about how "competition among managers is good." To be good, however, it must be healthy competition for *the same goals.* Too many companies still have Sales Managers striving for increased sales volume without regard to cost and profits, Manufacturing Managers keeping costs down regardless of the effect on customer service, Engineering Managers developing new product designs which cannot be produced and marketed at a profit, and other ridiculous examples of strong managers competing for different goals. What's needed is to identify and define the proper goals for the areas in which these major activities interact.

A good example is forecasting, an activity which is supposed to provide the information on customer demand basic to all planning. Attempts to assign sole responsibility for making forecasts to Marketing, however, have resulted in friction, criticism, buckpassing and frustration. If the marketing department makes the forecast they can always be criticized because it is wrong. Forecast errors provide excellent excuses for excess materials, missed delivery schedules and high production costs. I have seen dramatic improvements result, however, when marketing and materials control groups are given joint responsibility for improving the accuracy of future forecasts and also for making better use of present inaccurate forecasts by developing more sensitive error measurements, reacting to errors faster, and communicating reasons for these deviations more clearly. *Shared, not unique, responsibility, and mutual, not individual, effort toward such common goals assure success.*

We can look for increasing problems among those groups concerned with the bill of materials and its various uses. The bill of materials (or "formulation" or "product specification" as it is called in various industries) has long had two uses: Engineering uses it to describe the product and Manufacturing uses it to describe how it is put together. With the ordering technique called material requirements planning, the bill of materials becomes a vital part of the materials planning system. Materials Control uses it as the structure on which materials planning is hung. These three uses frequently require three different forms of the bill of materials. The question, "Who is responsible for the accuracy of each of these?" is not easy to answer.

One alternative is to have Product Design Engineering prepare the engineering bill, Manufacturing Engineering prepare the manufacturing bill and Materials Control prepare the planning bill. This results in problems of poor communications, delays in implementing changes and differences among the three bills of material. The best solution might be to define the basic role of the Product Engineering Department as not only to *design* the product so that it functions effectively, can be manufactured economically, and marketed competitively, but also to *communicate* via bills of material and product specifications information on product structure *in the form* needed by all who use it. I'm just not optimistic that the problem will be solved by forcing Engineering to maintain a file of bills of material structured in ways they don't understand, have no interest in and don't use. With computers and intelligent coding techniques, it is usually possible to serve everyone's needs with *one file*. This should be the primary objective of all groups. Joint responsibility for keeping this file accurate and up-to-date can then be assigned to all three groups—each responsible for its own inputs. The alternative is what most companies have now: three bills of material files, all different, all wrong.

As long as I can remember, Purchasing and Materials Control indulge in finger pointing and buck-passing whenever late deliveries of purchased materials are discussed. "Materials Management" was supposed to solve this problem too, but it didn't. Having the heads of these two departments report to one man will never solve the basic problem. What's needed is a definition of the primary responsibility of each group. Materials Control should plan material requirements and keep these plans up-to-date; there's not much argument on that. Ask any Purchasing manager what his primary role is, however; almost every time he'll say it's the "Six Rights"—getting the *Right* material of the *Right* quality in the *Right* quantity at the *Right* time and the *Right* cost from the *Right* vendor. Think about these six rights. Engineering specifies what the right material is; Engineering and/or Quality Control defines the right quality. The right quantity and the right time is determined by Materials Control. Purchasing simply relays this information to the vendor; all they contribute are more fingerprints on the paperwork. The only "rights" to which they make a real contribution are the price and the vendor selection.

*Their primary function is to buy materials at better prices from better vendors.* Why then do most Purchasing Departments spend about 85% of their time on clerical work, maintaining records which duplicate files kept by others, writing multi-copy purchase orders, checking receipts and invoices, playing middleman in follow-up and expediting of vendors? Only a small fraction of their time is spent on their primary function, even though purchased materials frequently exceed 50% of the total manufacturing costs. Why not free the Purchasing Department to concentrate on its real job? Let the computer maintain the files and produce the paperwork. Make Materials Control responsible for advising Purchasing well in advance of the dates orders must be placed with vendors so that adequate time is available to negotiate better purchase contracts. Good ordering systems will show which items will be ordered four, five or more weeks in advance. Let Quality Control handle directly with vendors questions of sub-standard material. Have Materials Control reschedule orders and expedite vendors; after all, they are the ones who really know when materials are needed. *Get the middlemen who serve no function except relaying information out of the act.* Let's abolish the absurd Purchasing motto, "Nobody talks to our vendors but us." They're not talking—just parroting other people's words!

The major objection to many departments dealing with vendors is that "too many people in the act will be getting our vendors all confused." This will only happen if you don't organize Materials Control, Quality Control and others who may talk to your vendors. There is no problem if one man (the buyer) negotiates price, terms and conditions, one (the scheduler) handles delivery dates and quantities, one (the quality expert) settles questions of good or bad, etc.

It is very revealing to look at purchasing in the "Systems Contracts" or "Stockless Purchasing" approach, widely applied to maintenance and repair items, office supplies and commercial hardware. Purchasing's role is to research sources, evaluate distributors and their product lines, negotiate prices, set up contract terms and conditions, establish criteria of good vendor performance and then get out of it! Actual orders are written and expedited by those authorized: the office supervisors, tool crib attendants, foremen and others who really know what they need. *Paperwork is not only simplified, purchasing is not involved in any routine matters.* What a contrast to the usual approach to buying manufacturing materials! The "fantasy of the six rights" should be converted into the "facts of the real objective"—better prices from better vendors.

There's also a great deal of work needed in redefining the roles and objectives of line Manufacturing. Almost invariably when customer delivery promises are not met or when stockouts occur, it's the Materials Control manager who is on the carpet. When should it be the Manufacturing manager? If their primary jobs are correctly identified, Materials Control should be held accountable only for poor planning; Manufacturing should answer for poor execution of the plan: not getting the required capacity to turn out the total amount of product needed or not working on the right items each day. In only a handful of companies, however, is it recognized that these two groups are complementary, not competitive, and their responsibilities placed correctly.

Foremen are usually held responsible for machine utilization, set-up time, employee downtime and overtime. They have about as much real control of these as you have of your teenagers. The level of machine and manpower utilization depends upon how well the Sales Department meets the forecast and how well the capacity was planned to begin with. Machine set-ups depend upon the lot-sizes issued by Materials Control and on orders being sent out in families so that they can be run together. Downtime and overtime depend on how well work is scheduled so that it flows instead of bunching up at bottlenecks.

*When foremen are measured on specific factors, however, they will do their best to control them; unfortunately, this usually compounds problems instead of solving them.* To meet set-up requirements, a foreman will hold orders until he can combine set-ups and run families together. This results in missing schedules. To improve machine or manpower utilization, a foreman will keep requesting more orders. Since downtime is such a heinous crime, he usually gets the orders. When it is finally recognized that he is too far ahead of schedule because he is over-manned, the pendulum swings in the opposite direction and slashing cuts are made.

Until the right primary goals have been established, no amount of organizational pressure, no sophisticated techniques, no crisis management can make things come out right. On the other hand, when primary goals have been identified correctly and mutual responsibilities set clearly, it hardly matters where individuals or work groups report in the organization structure. First, you have to know, "Where are we going?" The next question is, "How are we going to get there?" This breaks down into, "What's got to be done?" and "How will we do it?" Only then is it time to consider "Who's going to do it?" *After* the jobs are defined, the specific form of organization can be set up to take advantage of the strengths and to compensate for the weaknesses of people available. The harness should be adjusted to fit the horses.

Manufacturing control system logic has now been well-defined, and the role of the computer recognized as helping to provide managers with more timely, accurate information so they can make better decisions. These decisions can then be fed back into other computer programs, restructuring more information for further decision-making. This modular, segmented approach to information systems holds real promise for the development of "management information systems" utilizing common data files to generate information for all phases of a business. Managers won't be replaced by machines making decisions. Their decision-making power will be multiplied because they won't have to wait so long for the facts or do without them.

It seems safe now to predict that manufacturing companies in the future will include a Control Information Department in their organizations. This group will operate the computer and its related equipment, be in charge of the necessary files of Sales, Engineering, Manufacturing and Financial data to be used for planning in each of these areas, operate the feedback systems which measure actual performance against plan, and generate exception reports highlighting significant deviations from plan for decision-making on remedial action by operating managers. Long, intermediate and short range planning will be included. This department will incorporate activities now scattered among corporate planning, sales forecasting and statistics, inventory management, production control, manufacturing scheduling, quality control, purchasing, cost accounting, industrial engineering, product engineering, budgeting, warehousing, and product distribution. The benefits will be enormous, not only from reduced duplication of activities but also from more timely, accurate information for decision-making by operating departments with clearer ideas of their primary objectives.

Basic principles, the techniques needed for sound control, the necessary computer hardware and a substantial amount of the software are all available. What's required now is the hard management work of organizing for control—not looking for techniques, formulas or other panaceas. The right approach is defining primary goals, assigning proper objectives, developing competent people, providing them with information and maintaining the necessary disciplines to achieve control of operations. This cannot be achieved quickly but it must come. The advantages of being ahead of competition in this development are already visible in a few companies like ITT, Black & Decker, Dodge Manufacturing and Twin Disc. They have moved well ahead of their competitors. *Solid control of manufacturing undoubtedly holds more hope for some companies and more certainty of disaster for others than any other management activity.* It involves the brains of the organization, the only factor making one company completely unique from its competitors.

# 4

# Forecasts

## Good or Bad, but Useful

Forecasting in business is like sex in society: we have to have it; we can't get along without it; everyone is doing it, one way or another, but nobody is sure he's doing it the best way. Certainly the fundamental importance of the forecast in planning and managing a business is recognized by almost every manager. Among the typical attitudes I run into, the most common ones are, "You can't forecast our business" and "What we really need is a good forecast." The first usually comes from the Sales Department. They see all the difficulties involved in making an accurate forecast. They have been getting unreliable figures from their customers and they see how their business is affected by a host of factors: what their competitors are doing, changes in the business cycle, quality and delivery problems, etc. They probably have been criticized many times for errors in previous forecasts, and monuments to their big mistakes are still in the warehouse. They seem to hope that ignoring the problem will make it go away.

"If we only had a good forecast" is the cry of manufacturing people, particularly those in Materials Control. They apparently want a perfect forecast so that all their plans could be firm. It's fortunate they'll never get one—the principal excuse for their failures to plan and control properly would then be gone. They also feel no responsibility for forecasts. They pretend to be victims of other peoples' mistakes.

The water is muddied even more. Marketing people can and do point with pride to the good forecasts they made of total sales, or sales of a whole product family. Manufacturing people, of course, deride these figures as useless for ordering specific materials and scheduling individual products.

The depth of the frustration this bickering begets is measured by the enthusiasm which is generated by new forecasting techniques. Unfortunately, none of these has yet paid off. It's reflected directly, also, in the humdrum, lackadaisical, "Well, if we have to . . . " attitude of people forced to submit a forecast, people who have little confidence in the numbers and less understanding of why they are needed.

The really pertinent question is one that many top management people are now asking: "Are we really making the best use of the forecasts we have?" This recognizes that "making a better forecast" and "making better use of a forecast" are two distinctly different problems. Efforts to learn to use imperfect figures better may pay off far more than spending a lot of effort and energy trying to improve the accuracy of forecasting. *Forecasting is and. always will be an art*, one in which a point of diminishing returns is quickly reached—where additional effort to improve the accuracy of the figures will cost more than it's worth. On the other hand, learning to use forecasts more intelligently, based on basic principles now recognized, is an effort which pays continuing dividends.

In reviewing my experience with many companies over the years, I have seen six major reasons (Figure 5) why forecasts fail to produce more useful results:

1/ *Individual Effort*—No individual or small group within a single department and no statistical technique can possibly know enough about all the factors affecting the future to be able to develop an adequate forecast. Such a person or group can, of course, analyze historical data and use some techniques to draw a few conclusions about possible future trends, cycles and random variations. *Evaluating how the future will differ from the*

Individual Effort

Unrealistic Expectations

Second Guessing

Conflicting Objectives

Forecasting Wrong Things

No Timely Tracking

**Figure 5** / Why Forecasts Fail.

*past, however, is another question.* Internal company activities, such as new product designs, product improvements, better customer service, and marketing promotions, plus external factors like competitors' activities, changing customer preferences, technological developments, and the state of the national economy require inputs from many people, both inside and outside the company. *More important, however, is the need for the forecast users to understand fully the thinking behind the forecast.* To use a forecast intelligently, you need more than the numbers written on a sheet of paper. You need to know as much as possible about the factors which were evaluated in arriving at the numbers. Such understanding is too limited when the forecast is an individual effort. Effective forecasting must be a group effort.

2/ *Unrealistic Expectations*—Hope springs eternal in the human breast for the "perfect forecast." This hope always seems to be associated, however, with the idea that someone else will have to make it. *Failure to realize perfection results in rejection of the forecast but not the hope.* I've often wondered what production controllers would use for their Number One Excuse if they got a really good forecast.

3/ *Second Guessing*—Faced with the reality of bad forecasts, most of us exercise the privilege of making our own guesses. We know why the other fellow's figures were wrong as surely as market analysts tell us today why yesterday's stock market went up or down. Unfortunately, we are as able to predict tomorrow's figures accurately as those same market analysts are to tell us what will happen to stock prices tomorrow. In addition, second-guessers rarely communicate their knowledge and beliefs so that those responsible for the official forecasts can use them to improve the forecast. *As a result, the whole system degenerates.*

4/ *Conflicting Objectives*—The major groups within a company's organization view forecasts differently. Sales wants an optimistic figure, a carrot to be held out in front of salesmen to get them to exert themselves to the maximum to achieve top quotas. Top management and financial people, of course, never want to overstate projected profits. The Manufacturing group tries to strike some happy medium between the optimism of Sales and the conservatism of Finance in estimating what they will be called on to produce. Each of these is a valid objective. Unfortunately, in too many cases the forecasts developed are inconsistent—Sales, Finance and Manufacturing use forecasts with significant differences in product mix as well as in the total figures. *The result is that the battle for company survival is fought out with different strategies being used by the major forces.* As if competition wasn't enough of a problem!

5/ *Forecasting the Wrong Things*—Figure 6 shows four End Products (or finished products) whose sales to customers are forecast at the weekly rates shown on line "D"—demand. Their inventories are replenished in the lot quantities indicated on line "P"—production. Making finished products in lots greater than one week's supply is very common, of course. In some cases it is done to reduce the cost of changing over the lines and in others because some of each product cannot be made every week. *Demand for so-called "independent demand" products must be forecast*; there is no other way to develop estimates of future requirements. For their subassemblies, however, there is no need to forecast the demand. It can be calculated using the bills of material showing which subassemblies and component parts make up each end product, and the lot sizes in which the end products, subassemblies and parts will be produced. *Any company making "assembled products," no matter whether screwed, glued, sewed, packaged, welded or chemically joined, can reduce dramatically the number of forecasts it needs by making them only for the end products and calculating requirements for all components of these assemblies.*

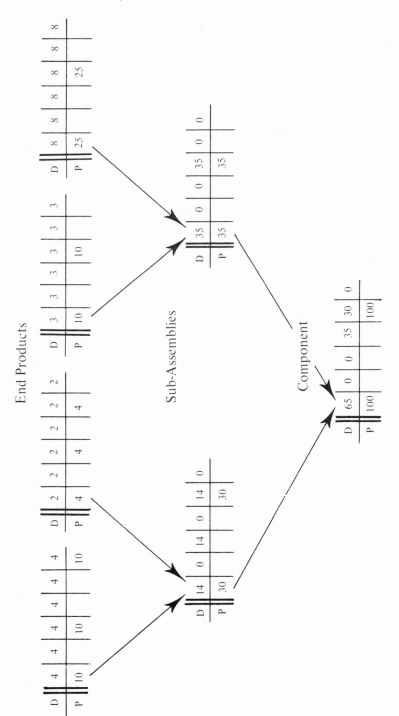

**Figure 6** / Forecast vs. Calculated Requirements.

Preoccupation with individual finished products blinds some companies to the right things to forecast. A forging manufacturer saw no way to make a forecast of the products ordered by his customers. Although they got some repeat orders, the bulk of their output was custom-made to suit the customer's drawings. They badly needed information on future requirements, however, to plan capacity—to be sure they had enough people running enough forging hammers on enough shifts. I asked them what they would do with the information if I had a magic slate which would show them specifically all the forgings their customers would order. Obviously, they would estimate the standard hours of work on the various sizes of hammers required to produce the specific forgings. *What they really needed, of course, was a forecast of this standard hour work content of their customers' orders and not of the individual piece parts.* Historical data updated by a simple averaging technique furnished them with some very useful capacity requirements planning information. Furthermore, they were able to "manage" this forecast. When they saw the load on some sizes of hammers falling off, they had their salesmen contact customers who customarily bought forgings produced on these hammer sizes, offering them immediate delivery or better prices to place orders now.

Many companies' products give their customers a wide variety of options in selecting features, accessories and alternatives in various combinations. The outstanding example is the automobile, where literally billions of different cars could result from the combinations of options customers might choose.

Figure 7 illustrates a fairly simple product, an electric hoist, manufactured in a family of sizes, Each hoist in the family consists of a motor, gear box, drum assembly, set of controls, and hook. The customer has the number of options indicated on each of these features, and the Engineering Department is working hard to give them more options on the hook. A maximum of 2400 combinations of options is now available to customers; they could order that many different hoists. As soon as Engineering designs another hook assembly, this will double to 4800. It is evident that *no technique, however sophisticated, and no individual, however much of a genius, could forecast how many of*

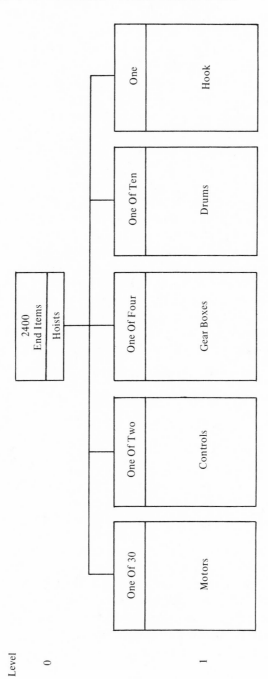

**Figure 7** / Forecasting Options.

*each of the possible combinations customers will order and even come reasonably close*—particularly when total sales of the family is only a few hundred a month. Insisting that a marketing department produce detailed forecasts of finished products like these is as likely to produce useful results as lighting matches in the open on a windy night. I've seen a marketing department go from antagonism to enthusiasm, however, when asked to forecast a total for the family, and a percentage of this total for each option, instead of forecasting individual hoists. Forecasting the right things can produce dramatically better results.

6/ *No Timely Tracking*—Many of those who have to work with forecasts produced by others feel like victims of their numbers. This is completely unnecessary. No one needs to be a victim of not knowing how wrong a forecast is likely to be. Anyone with access to actual sales data can compare them to forecasts and see how wrong they have been in the past. Surprisingly enough, using past deviations as a forecast of future error turns out to be a lot more accurate than using past demand as a forecast of future requirements. And this error holds the key to better use of bad forecasts. *There are two key words in tracking forecast errors—do it REGULARLY and do it OFTEN.*

Forecasts have five basic characteristics, listed in Figure 8:

Forecasts are:

    Always wrong.
    Two numbers.
    More accurate for families.
    Less accurate far out.
    No substitute for calculated demand.

**Figure 8 / Forecast Characteristics.**

1/ *Forecasts are always wrong*—Everyone recognizes that this is a fact—forecasts *will* always be wrong—but it is surprising how few people ever plan on the forecast being wrong and get ready to react to the errors that will inevitably result. Apparently history teaches them nothing; they keep hoping that accuracy, like peace, will come because we all wish for it.

2/ *Forecasts should be two numbers*—Since forecasts will be wrong, it's vital to have some estimate (itself a forecast) of how wrong. Any forecast that doesn't indicate a range ("$\pm$ 15%", or "2300/2800") is only half a forecast. There's no excuse for lacking some measure of accuracy. Even those not involved in making the forecast can track actual sales and develop data on past performance.

3/ *Forecasts are more accurate for families of items*—Few companies have errors of more than five percent in their forecasts of total annual sales. If they could forecast individual products nearly as accurately, they'd have far fewer troubles. Wherever it can be done and still suit the purpose, forecasts should be made for *groups* of products. This will be possible in equipment and manpower planning, ordering common components, buying capacity from vendors, and many more cases.

4/ *Forecasts are less accurate far in the future*—There's frequently a strong feeling that far future forecasts couldn't be any worse than next week's forecast, for example. Many see clearly that actual demand in specific periods never matches the forecast. Few understand that, at best, the forecast is the *average* quantity to be expected in each of a given number of periods. The forecast of average sales next week will be more accurate than one for the corresponding week next year. A forecast for the last quarter of this year will be better than one for the same quarter next year. *This is the strongest reason for short lead times.* When you're planning materials for specific products you can't get any help from the third characteristic. For any hope of minimizing forecast errors, you need short lead times.

5/ *Forecasts are no substitute for calculated demand*—Customer demand for finished products must be forecast if information is needed now to meet requirements in time periods beyond those covered by the backlog of unshipped customer orders. There's no other way to get it. Demand for *components* of these products (assemblies, subassemblies, parts, raw materials, etc.) can be calculated, however, as illustrated in Figure 6, and need not be forecast. Master production schedules can be developed indicating how many of each product will be assembled, and when, to meet the forecast demand. A component's demand then obviously depends on which products it's used in, how many are already on hand or on order, and how many will be made or bought in one lot—simple calculations based on bills of material and inventory records. The very great benefits of this approach, called material requirements planning, compared to attempting to forecast each component's demand independently, are covered in Chapter 5.

Forecasts must and will be made. If a formal, official forecast is not prepared, everyone making decisions about future activities—ordering materials, hiring people, buying equipment—will be forced to make his own estimates. It would seem evident that it couldn't help but be better if the best-informed people were involved in a formal effort to produce an official forecast which would provide a consistent basis for all planning activities. Don Gates, formerly Director of Marketing Research at Dodge Manufacturing Company, recognized this and set up a five-step program, shown in Figure 9. This was not a one-shot effort but rather a continuing program to make and use forecasts more intelligently. It is well worth considering, regardless of the type of business you're in.

1.  Define the purpose of each forecast.

2.  Collect and analyze historical data.

3.  Develop and refine a forecast model.

4.  Evaluate internal factors.

5.  Evaluate external factors.

**Figure 9** / Forecast Program.

## DEFINE THE PURPOSE

Forecasts have many well-recognized purposes, of course, and they need no repetition here. It is interesting, however, to view the approach many companies take, developing separate forecasts for sales purposes, manufacturing needs, and financial uses. These were discussed earlier in this chapter. This approach resembles an army on the march whose generals have one destination, the quarter-master corps another, while the main body of troops is heading somewhere in between. This would be bad enough if the figures had a common base and didn't spring from three separate sources, but frequently each group makes its own forecast—three distinct strategies! What a way to fight a war.

Don Gates' approach, defining the purpose, permits us to distinguish among such different purposes for Sales, Financial, and Manufacturing operations, but I can't overemphasize the need for a consistent base from which to develop these. Once such a base is established, Sales figures can be increased to push out the carrot for the sales effort and Financial and Manufacturing plans can be linked closely together. How else can you get everyone fighting the same battle?

Defining the purpose also sets the length of the forecast period. A forecast for capital investment purposes would be long range, two to five years. At the other extreme, forecasts for procuring purchased materials and scheduling production would cover next week or next month. Once the forecast period is established, the frequency of review is also indicated. It is not necessary to revise an annual forecast more than once a month; on the other hand weekly forecasts should be updated at least weekly.

Setting the purpose of the forecast also establishes the units in which it should be expressed. Profit plans and sales quotas are only useful if given in dollars, while manufacturing data must be expressed in pieces, tons, gallons or similar units meaningful to people establishing plant capacity or ordering material. The important requirement is that all forecasts be consistent so that the army and all its related services are marching toward the same goal.

This approach will frequently show how to make a forecast when it might not be clear that one is even possible. The forging manufacturer discussed earlier in this chapter is an excellent example. Their thinking was locked up with forecasting specific parts. However, their purpose in forecasting was to establish equipment and manpower capacities; they needed a forecast of forging hammer hours, not specific pieces of individual forgings. Their demand history showed that it was possible to use statistical methods to project future requirements from historical data on the work content of past orders. They were able to make a valid, useful forecast for developing manpower and forging hammer needs.

When the purpose of the forecast is to provide input for planning material requirements, the type of forecast needed will be greatly influenced by product structure and options available to customers. This will be covered in detail in Chapter 5.

## COLLECT AND ANALYZE HISTORICAL DATA

The purpose here is to see if anything can be learned from the history of previous demand. Since the intent is to project into the future the trends and cyclical tendencies exhibited in the past, historical data should be purged of any one-time occurrences which might not reoccur in the future. Big government orders, material to fill a new distributor's pipeline, peaks of demand due to pre-announced price changes and the like have to be removed if valid conclusions about the future are to be drawn from past history. A tough problem is converting the data from the units used in existing available records into those needed for the forecast. Analyzing and purifying such data takes a thorough knowledge of past activities both in the marketplace and in the company.

Statistical historical information should also be separated into "streams of demand." Spare parts requirements should be treated separately from assembly needs. Products sold in high volumes to original equipment manufacturers should be treated differently from the same items marketed through company-owned distribution channels. Demand for carload quantities should be separated from demand for small lots. This separation is a vital requirement if the patterns of past history are to be interpreted correctly.

## DEVELOP AND REFINE A FORECAST MODEL

The objective is to find one or more usable statistical forecasting techniques. These divide into two basic classes: those using the item's own demand history to predict future demand and those using historical data for related products to develop an item's forecast. These may be as simple as an arithmetical average or as complex as regression analysis and multiple correlation. These techniques are well covered in the literature. Simple or complex, any "model" is only "a small imitation of the real thing." Don't expect any mathematical formula to handle the complexities of your business without your people being involved. *All statistical approaches make the basic assumption that the future will continue to be like the past.* Unfortunately, the impression gained from proponents of these techniques is that all you have to do is to put these techniques on your computer and your forecasting problems will be solved. This has not happened yet and never will.

Before any attempt is made to use statistical techniques, they should be thoroughly tested. Simulations should be run using historical data, assuming the technique had been in use a year or two ago, and seeing how well it would have predicted the actual demand that developed. For strongly seasonal demand, you may need three years' data. Such simulations usually aim to select a model which will minimize forecast error, and statisticians have some precise methods for measuring this error. Such analyses lead the sophisticated specialist to search for more and more complex models which "do a better job of tracking erratic demand patterns" as proved by smaller errors. A more practical approach is to measure the forecast model against the *average* of the actual demand. This average is what you are really attempting to forecast. If the average can be approximated closely by a simple model there is no need and no justification for getting involved with fancier mathematics.

Software packages, available from many sources, embody the more popular mathematical forecasting techniques such as simple exponential smoothing, with trend or seasonal adjustments. Some of these programs will even select a "horizontal" or a "trend" model for projection of the forecast, based on the characteristics of the demand reported to them. Others will adjust automatically the weighting factors for weighted-averaging techniques as the actual demand patterns change. It's dangerous to allow such automatic adjustment of forecasts to occur without some human judgment being exercised. Such sophistication is well past the point of diminishing returns in arriving at a forecast which is valid, reasonably accurate and, more important, well understood by those using it. My associate, Ernie Theisen, summed it up beautifully. The word "forecast," he said, "has two parts—'fore,' from golf, meaning 'look out,' and 'cast,' from fishing, meaning 'throw out'." This is exactly the right way to view statistical forecasts. Watch them—then discard them when invalid.

## EVALUATE INTERNAL FACTORS

In many companies I'm familiar with, some mathematical model was found which could be used to make a reasonable projection of demand, as long as the future continued to be somewhat like the past. For a sound forecast, however, specific evaluations must be made of those factors which will make the future different from the past. New product designs, marketing promotions, price changes, quality improvements and lead-time reductions are all aimed at making the future better than the past and increasing customer demand. Estimates of their influence must be converted into a quantitative adjustment of a statistical forecast. Obviously such judgments should not be left to clerks or computer programs.

## EVALUATE EXTERNAL FACTORS

Changes in the business cycle, activities of competitors, the tightness of investment capital and many other factors external to a company's activities must also be evaluated if they are likely to have a significant effect on the demand for its products. Such factors have a greater impact for long-range forecasting to plan capital investment in plant and equipment than for the day-by-day inventory decisions. Their influence will be felt more on product families than individual items and be much more difficult to express in quantitative terms. Here again, management judgment is required and even rough attempts at such evaluations will be better than no attempt at all.

It is obvious that computers and mathematics by themselves will never make adequate forecasts for any business. A combination of statistical analysis to project trends and cyclical tendencies and to provide a base on which to apply management judgment seems to be the soundest, most rational approach. It becomes evident when looking at the five steps in Don Gates' program that this cannot be handled by any one department or group within a company. Marketing and Sales obviously play a major role, as does top-level management, in evaluating the internal and external factors. Significant contributions can be made by the Materials Control group and by Data Processing in collecting and analyzing data, developing and refining forecast models, and operating routine systems to update the forecast. The Financial and Cost Accounting people can frequently add something also because of their training and ability to study historical data and analyze its characteristics.

The obvious conclusion must be that no mechanical means will be adequate for forecasting; rather, the combination of a mathematical base and the judgment of people "in the know" is required to develop the best estimates of what the future will bring. A technique which has worked well in many companies is to have a computer using a statistical technique develop forecasts for individual products. These individual forecasts are then totaled into family groups. (Family groups can also be forecast as if each were an individual item and these totals compared to the sum of the individual product forecasts). The group totals arrived at are then reviewed by marketing and top management people who decide if overrides or alterations are necessary. Any changes are then prorated among individual products in the family to develop each item's forecast. This method uses the power of the computer to update many individual forecasts frequently. With this approach, management judgment needs to be applied only to a limited number of product families. Management can then take adequate time to review these thoroughly, thinking about internal and external influences which might significantly alter the statistical data.

Econometric models are apparently becoming more and more useful. These try to measure the impact of one economic variable (like price increases in basic metals) on another (like production of household appliances ) in order to predict future developments in industry and the economy. They consist of a series of equations supposed to simulate the economic process. Using these equations, and assuming values for a few variables, the model calculates the others.

To date, the best contribution of econometric models has been in developing total industry data. They have failed to tip off changes from growth to recession, upward to downward trends and vice versa. These are the hardest kinds of change to predict, however. Their usefulness is expanding as more data are accumulated relating what one industry buys from and sells to others. The groupings are becoming sufficiently definitive that meaningful information can be obtained for some specific product lines. While they may become more useful for long-range financial planning, capital investment forecasting, and long-range capacity planning, it is obvious that econometric models will never really make a contribution to forecasting individual products. They will certainly be helpful to management, however, in evaluating the influence of external factors on their business.

New products are among the most difficult to forecast. If any historical information is available, it's usually too skimpy to use to make projections. Human judgment is about the only approach.

Equally difficult to predict are style items, special promotions and specific marketing programs. Exceptions to this can be made in some style items in the clothing and fashion businesses, and in products like collectors' coins, and toys which have a very short but fairly well-defined life cycle. For such products, historical data on similar items or programs can be used to develop characteristic life cycle curves similar to Figure 10. A range of possible error could be indicated by drawing an "envelope" with the curve as shown. New products expected to have the same life cycle can then be forecast and the forecast adjusted from actual data as it develops. Using the curve in Figure 10, a product estimated to be at the 40% point in its life cycle should have achieved between 10 and 20% of its peak cumulative sales demand. Applied to the actual sales experienced, these percentage figures indicate the approximate total to be sold throughout the life of the item. Two major problems are present in this approach; first, to determine whether or not a well-defined life cycle curve exists and second, to estimate at what point in the life cycle a particular product may be at any moment. The approach has worked well, however, for many products where these conditions can be met.

For new technological developments, an approach known as the "Delphi Technique" has been tried, and is claimed to give some benefits. A panel of "experts" is asked to make forecasts—anonymously—of the effects of some technological development such as "cold light" sources. Each one then gets a summary of the others' forecasts and makes another estimate. This is repeated until the guesses stabilize. Typically it's been applied to forecasting the future of new commercial aircraft, drugs, building materials, etc. Although it has yet to be really tested in industry, it would appear that the Delphi technique might have some contribution to make in getting a weighted consensus from a management team on a product line or process development which is innovative and unique.

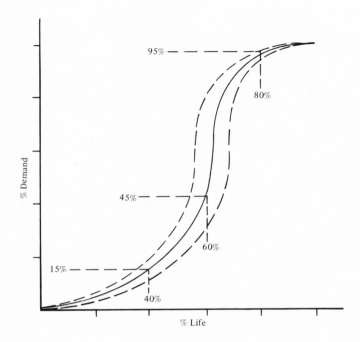

**Figure 10** / Life-Cycle Curve for Forecasting.

## USING FORECASTS

Using forecasts intelligently holds much more promise for progress than attempting to make better forecasts. The first two principles of forecasting are the keys. The recognition that forecasts will be wrong and the measurement of error as actual data are accumulated to compare to forecasts provide the bases for reacting to change.

Figure 11 illustrates two tracking signals which can give early warning that forecasts are not performing properly. The first is a simple comparison between the last period's actual sales and the long-term average, say six months or a year. This is the simplest and crudest approach possible but, like many crude tools, it can be

$$T.S. \quad = \quad \frac{\text{Last Period Actual}}{\text{Long Term Average}}$$

$$T.S. \quad = \quad \frac{\text{Running Sum Errors}}{\text{Average Error}}$$

**Figure 11** / Tracking Signals.

remarkably effective in the hands of an experienced person. The McGill Bearing Company in Valparaiso, Indiana, uses manually prepared charts similar to Figure 12 for "A" items—their highest selling bearings. They chart actual sales by month as compared to the forecast. At very little expense, these charts provide useful early warnings that major products are not performing as expected.

The other tracking signal in Figure 11 develops an index which can be used effectively for many items. For a large number of products, a computer will be required to handle the data with this tracking signal. It develops a running sum of the forecast errors, adding the deviations when actual sales are above forecast and subtracting them when they are below, and divides this running sum by the average forecast error. If the forecast is performing properly, plus errors and minus errors will tend to cancel each other out and the tracking signal will stay small. A forecast consistently low or high will develop a steadily increasing running sum; this will cause the tracking signal to grow steadily. Some specific value of the tracking signal (four or five, for example) can be selected so that the system calls attention by exception to those forecasts which are not performing well. Dividing the running sum of the errors by the average error prevents a highly variable demand (which would produce large error sums) from causing false alarms when the forecast is accurately predicting the average.

Too little thought has been given, however, to applying forecast error to planning for production tooling, dies, jigs and fixtures as well as material items. No matter how large the anticipated error, it can be useful in applying judgment to planning the procurement of such items. Here's an example. A company introduced a new

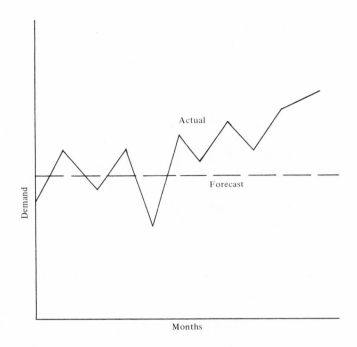

**Figure 12** / Actual vs. Forecast Chart.

product which was similar to others in its line but was still sufficiently unique to raise serious questions about the forecast. A preliminary figure of 40,000 per week was proposed by Marketing. Pinned down about possible variations from this figure, they said their most optimistic figure was 70% higher and their most pessimistic 70% lower—in other words, a range from 12,000 to 68,000 per week. The first reaction was, "ridiculous." To which Marketing replied, "You asked for our best estimate and you got it. What number would you like?" This put everyone's viewpoint back into perspective and they settled down to see if even such a large error measure could be useful in making some critical production decisions.

The number of forging dies to make for the first production runs, for example, could be determined by the low forecast. Extra dies can be made by EDM (electrical discharge machining) from carbon masters in short lead times and at relatively little expense compared to the time and cost of the first sets of dies. On the other hand, multiple cavity plastic molding dies could be designed for the high forecast, since the incremental cost of additional cavities in the first die is relatively small compared to duplicating dies later to achieve the same total production capacity. In the case of both forgings and plastic moldings, parts could be produced based on the middle forecast so that adequate supplies would be on hand to get the product off and running, fill the pipeline in the distribution system, and support sales if they should"take off." For castings, ordering wood pattern equipment with limited productive life would be smart until actual sales had justified the investment in high-output, metal match-plate equipment. Jigs and fixtures to hold parts for turning, boring, milling and drilling could also be simple and inexpensive until the actual sales rate was high enough to justify the higher investment in more specialized, high-rate production tooling. Conversely, small inexpensive items such as commercial hardware and packaging materials could be ordered to the high forecast to be sure that lack of such items would never hold up the program. Any loss due to overstocking of these inexpensive materials would be a relatively small price to pay for the insurance against holding up shipments of a successful program and for the freedom to give full attention to the important few components at a critical time. An interesting footnote to this case is that the product really took off but production stayed with it. They had hedged their bets smartly and watched actual sales closely.

The estimate of error is obviously at least as important as the forecast of demand. Unfortunately, too few companies measure and estimate error; this is vitally necessary for good use of forecasts. The error will never be eliminated. The only alternative is to use it intelligently.

The third and fourth principles of forecasting also hold keys to using forecasts better. For capacity planning, manload determination, and similar intermediate and long range forecasting, groups of items rather than individual products should be used. The best approach appears to be to forecast the group of products as a single item, rather than forecasting individual items and adding them up to get a family total. These group figures can be plotted cumulatively on a chart like Figure 13 to indicate when

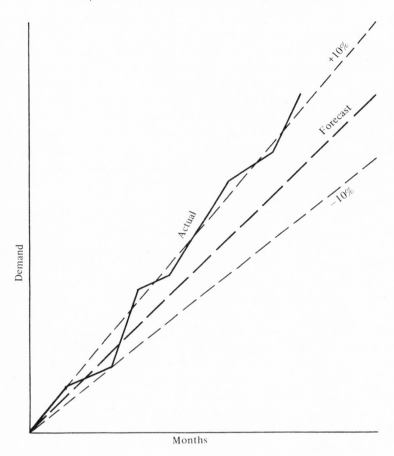

**Figure 13** / Cumulative Actual vs. Forecast Chart.

forecast revisions and changes in plan are needed. More on this will be found in Chapter 6.

Materials control involves forecasting individual items, and I've already said that lead times must be kept short so that forecast error is reduced. Short lead times are desirable since they minimize variations in demand and changes in planning. Much more will be said on this point also in Chapter 6. Many forecasting problems, together with excess inventory, shortages of components, and crises in the plant could be *eliminated* (literally) if materials control people *didn't forecast* but instead *calculated* requirements for parts being assembled. This technique, called Material Requirements Planning, will be discussed in Chaper 5.

Forecasting is remarkably like auto safety; it's a serious problem to the people involved—the drivers and pedestrians in the case of automobiles and the sales and inventory control people with forecasts. In both fields we have a few experts whose names are familiar. These men are mounting a massive assault on 5% of the problem. We want and need better, safer automobiles and better, more accurate forecasts. Unfortunately, 95% of the disasters result from improper use of both. I'm not saying we need less emphasis on safer cars and better forecasts. I am saying we need a lot more emphasis on intelligent use of both.

Defensive driving is based on the principle of expecting the worst and being prepared to react to it. This is exactly what we need to do to get better use from forecasts. We need more and better early warning systems. Our DEW line should be *DON'T ENJOY WAITING* until your plan is obviously faulty. Expect it to go wrong, make plans now for what you will do when it does develop significant errors, and take action when you are convinced the original forecast is no longer valid.

Forecasting has been and still is a "black art." Fortunately, we've outgrown the practices of studying chicken entrails, subjecting ourselves to drugged trances, and gazing into crystal balls to divine the future. We've got some good tools: statistical techniques, tracking signals, computers among them. Let's give up the pious hope of finding Paradise, roll up our sleeves and get down to the tiring, frustrating drudgery of learning how to plan for, detect, and react to change. Change is inevitable and isn't it what we all really want?

# 5

# Planning Inventories to Suit Your Resources

Despite the importance of proper planning and tight control of inventories to a company's economic health, few companies really do a good job. The obvious symptoms are high inventory levels relative to sales rates, poor service to customers, excess costs, inventory budgets rarely met, and large annual writeoffs every two or three years. Is it really as tough a job as this would indicate? Certainly poor performance isn't caused by lack of desire on the part of management to get inventories under control. Few managers are unaware of the amount of capital tied up in inventories, or the need for improving turnover.

Most managers use turnover as their measure of control of inventories, probably because they have no other measures available. It has two major deficiencies. While it is a good average measure of *progress in improving control* of inventories, it gives no indication of what is "par for the course." Don't look up the turnover rate of a competitor or a "similar" business and compare it to yours. Such comparisons are valid only if your product line is the same, if they add as much value to the materials bought as you do, using essentially the same processes, plant and equipment, and if they want to run their business as you run yours, having similar employment policies, distribution systems and sales and pricing policies. Obviously these are not the same for many companies. Neither should turnover be used as an inventory planning technique to decide when and how much to order, since it ignores potential economies in such decisions.

Is inventory an asset or a liability? On all balance sheets it is listed under "Assets" but here the resemblance often stops. Financial people exert continual pressure to reduce inventories; listening to some you'd think the right turnover rate was infinite. Many manufacturing people get fidgety and upset whenever bare spots appear on the plant floor; they seem to need plenty of inventory sitting around unused in a full plant, although they can't tolerate low machine utilization. Sales people deplore the "inventory" of unshipped orders but shrug off excess finished product as "good insurance" or "just one of those things."

Most companies have capital appropriations over $5,000 reviewed by high-level managers, but inventory clerks can order $100,000 worth of materials when they think it necessary. When do we make inventory justify its existence by earning a return on the investment? Few managers can even say specifically *how* it should earn any return, yet such justification should underly all inventory planning.

Inventory planning involves answering four basic questions:

1/ What to control and how tightly?

2/ How much to order?

3/ When to reorder?

4/ Should total inventory decrease or increase?

You rarely have enough resources of money, manpower, or time to plan and control inventories as you would like to. For this reason, your attention must be focused on the most important items so that available resources can be applied to control them most effectively. Simplified techniques can then be applied to the balance to maintain at least adequate control. This requires classifying inventories.

Pareto's concept distinguishing the "vital few" from the "trivial many" really applies to inventories. Whether you're thinking of materials now in stock or the dollars you spend over a year, a very few items make up the great bulk of the total value. The most familiar technique ranks items in inventory according to the *annual usage value*, not the dollar investment in inventory at any particular moment. The "vital few" are those items which have the highest dollar values flowing through the inventory; these obviously should receive the bulk of planning and control attention. Companies with highly engineered products such as medical research equipment, heavy machinery and electronic devices will have a very skewed distribution, with about 5% of the items accounting for over 90% of the annual usage value. The closer the company gets to the ultimate users of its products, the closer these figures come together. I have rarely seen an example, however, where more than 25% of the items constitute less than 80% of the annual usage value. In all cases, the "A" (high-value) items can be split up also, and it will be found that a small percentage of these comprise a very high percentage of the annual usage value. Making such an analysis is the first step in good inventory planning and control: identify the important few items, focus attention on them, get tight, effective control, but *don't ignore the rest.*

Materials control books and articles contain many examples of simplified techniques to use for the low value "C" items which require no records or a very minimum of recordkeeping. The principle is, "Have plenty of these items; avoid the risk of stockout because this is cheap insurance." The intent is to *use the time and effort saved on such items to do a better job of controlling the important "A" and "B" items for better control of the total investment.*

Whatever technique is used, some discipline is required to make it work. This is where most companies get into trouble. Disciplines are relaxed because of the general impression that "we have plenty of such items so why worry about them?" As a result, shortages of bolts, nuts, washers, and similar trivia cause delays in assemblies and missed delivery dates. The few disciplines required with even the simplest systems—good physical control of the stocks, following the established procedures, and reviewing annual requirements—must be maintained if even the simplest technique is to function effectively. On the other hand, a high degree of precision in making the classification is not required. There's no need to include every item in the analysis or to use the latest, most accurate standard costs. The analysis should be made periodically, using the best data available at the time but without hanging up on unnecessary precision. The purpose is to get guidelines to focus attention so that scarce assets are best employed.

More important is the need to decide how to group materials when making the classification. Obviously, if all inventoried items are grouped when making the classification, the high-volume finished products and their subassemblies will dominate the "A" group. It may be more important to identify the "A" items in each product group, or to distinguish between assemblies and subassemblies in one case, or between purchased and manufactured parts in another.

A-B-C analyses are made for a variety of purposes, and the purpose should determine how the items are grouped. For example, if different planners order and control different product lines, each should know his "vital few" items, and separate classifications for each planner's products would be made. Many items are subject to obsolescence because of engineering changes; it might be helpful to avoid large orders and excesses by ranking such items according to the probability of an engineering change in a given time period, thus identifying the high potential dollar excesses.

There may also be exceptions to the results of routine A-B-C analyses. While its annual usage value may be low, the unit value of an item may be high; a $500 diesel engine which showed up as a "C" item because only a few are used would certainly not be controlled by a two-bin system. The point to remember is that the purpose of classifying is to focus attention on those items where the greatest control can be achieved while making best use of scarce resources available in time, manpower and money.

Another classification that should be made is Active—Slow Moving—Obsolete. Here the basic purpose is to detect a change in the rate of usage soon enough to avoid having large quantities of obsolete material on hand. W. Evert Welch of ITT once stated, "You have obsolete inventory because you ordered too much the last time you ordered." While this may seem an extremely naive statement, it is well worth thinking about. What Evert Welch was really saying was, have some means to detect a falloff in usage rates soon enough to avoid obsolete inventory. He was also saying that there's nothing economical about ordering material you will not use, regardless of what formulas and calculations might indicate.

Some items in almost every product line are much more prone to obsolescence than others, and these should be reviewed before reordering to see that no design changes are contemplated which would make them useless. Engineering departments should know which are the "A" items in inventory so that they can alert production control when they are contemplating a design change on such items.

There are four basic causes of obsolete inventory:

1/ Customer apathy. Either a mature product has reached the end of its sales life or a new one has little appeal to customers.

2/ Poor planning and control systems. Too insensitive to detect drop-off in demand in time to react, or too undisciplined to be sure the rules are followed.

3/ Record errors. True quantities on hand are larger than the record balances and excesses are not worked off, even if it's planned properly.

4/ Poorly controlled engineering design changes. If they were planned properly, remaining stocks could be balanced out and worked off.

The penalties for being left with dead inventory includes the wasted capacity used to make it, the cost of carrying it until it's disposed of, and the cost of reworking it or writing it off. Clever ways of disposing of obsolete material are no substitute for simple, well-managed techniques for preventing it.

Blanket orders for "A" items should be hedged with cancellation clauses in the event of design changes. More will be given on this subject in the Capacity Planning and Control Chapter, where buying capacity instead of specific items will be discussed. Inventory records should contain some measures of rates of usage such as average monthly or weekly issues. These can be compared regularly to forecasted rates to detect a significant dropoff in consumption so that prompt action can be taken to minimize the amount of obsolete inventory. It's far better to make a considerable effort to avoid obsolescence than to develop ingenious (and frequently expensive) methods of utilizing obsolete material.

Other methods of classifying inventory are often helpful. For example, an A-B-C analysis based on the actual dollar value of inventory of each item now in stock would indicate items with large dollar investments which would have to be reduced to affect the total significantly in the short run. Grouping items in terms of time periods of supply (weeks or days supply on hand) can also be useful to focus attention on excess inventories. Having a record of the date of the last transaction on items highlights those where activity is slowing down. A little ingenuity in classification will do much toward achieving maximum benefits with minimum effort.

## HOW MUCH TO ORDER?

The concept of the "economic ordering quantity" is one of the oldest in the field. It was first published in 1915 but has yet to achieve anything like its full potential. It applies to items which are replenished periodically into inventory in lots covering several periods' needs or those where quantities required in several periods can be combined. The idea is to order that quantity which gives the *minimum total of the ordering costs and the inventory carrying costs.* Obviously, the larger the quantity the greater will be the investment in inventory and the higher the costs of carrying this inventory; at the same time, costs associated with ordering the item will be lower. For each item and its planned usage over a long period, there will be some ordering quantity which minimizes the sum of these two groups of costs. The literature of the field is full of formulas and their variations to cover a wide variety of conditions. The oldest and most familiar is called the "Square Root EOQ" or the "Camp Formula."

The basic assumptions underlying all of these formulas are that the usage rate will be uniform, that all items ordered will be used up, that only the costs included in the formula are significant, and that the plant or vendor can handle the order. Where these assumptions are valid, proper use of the formulas can make significant benefits possible.

Unfortunately there have been relatively few who have realized these significant benefits. Faith in the concept and its ability to achieve savings has led many to search for the proper formula, find the costs and other data needed, put these numbers blindly into the equation, and assume that the results will be beneficial. They almost never are. *Savings are not generated by formulas but by people taking effective action.* Economic lot sizes can yield benefits only if inventories are really reduced or if less money is spent on ordering or setting up machines. This will be discussed more thoroughly in the section "Total Inventory Management" to follow.

The proper costs to use in economic lot-size formulas have been the subjects of much debate. To achieve real benefits, only out-of-pocket costs which are influenced by changing lot-sizes should be included. Costs in standard accounting systems cannot be used since they invariably contain fixed elements.

Several basic assumptions of the square root formulas for economic ordering quantities are obviously not valid for items used as components of assemblies or raw material used to make a manufactured part. The usage rate is not uniform, the large and lumpy quantities used will result in small residues from the original order quantity which will not be used up, and the plant or vendor may not be able to respond in the variable periods at which orders are released. To handle assembly components, raw materials and other so-called "dependent demand" items, discrete lot-sizing techniques have been developed which overcome these difficulties. A variety of such techniques is now available to use when net requirements for an item can be determined for each of several future time periods and these net requirements used in place of an overall total requirement for a year to calculate order quantities. The literature of the field has details on these, so practitioners have little excuse for not at least knowing about them.

In using any of the economic ordering quantity techniques, professional practitioners know better than to use formulas blindly without investigating the practical implications and real-life effects of applying the numbers generated by the techniques. In many cases tool life, available machine capacity, space limitations, capital shortages and other limitations make the calculations invalid. More will be discussed on this subject also under Total Inventory Management in this Chapter.

## WHEN TO REORDER

Probably the most important of the three basic questions of inventory planning is the question of when to place a replenishment order. If it is issued too soon there will be excessive inventory on hand when the new lot is received; if it is issued too late there will frequently be a stockout, with all of its accompanying ill-effects. There are basically only two techniques for determining when to place a replenishment order. These are Order Point/Order Quantity and Material Requirements Planning.

### Order Point/Order Quantity Techniques

Order point techniques include such well-known approaches as Min/Max, Hi/Low, Periodic Review, and all of the simplified "C" stock control tools, as well as the more sophisticated statistical analyses. They attempt to plan a replenishment time by forecasting the average rate of demand during the standard lead time, and calculating a "Reorder Point" or an "Order Review Point" including both this expected demand and a "safety stock." Since it is recognized that demand forecasts are inaccurate and lead times vary, additional inventory called reserve stock or safety stock is added to the order quantity to guard against such variations. The literature of inventory control is full of higher mathematical techniques for setting reorder points and dynamic means for recalculating these as forecasts are updated and demand variations change. All attempt to determine the specific amount of safety stock needed to assure a definite level of customer service. Techniques vary for different ways of expressing "customer service level."

The basic assumptions underlying this group of techniques are:

1/ Usage will be relatively uniform and the forecast average will be representative of demand that will actually occur.

2/ Forecast errors in future periods will at worst be no larger than in the past.

3/ Usage will be in quantities relatively small compared to the replenishment or reorder quantity.

4/ It is desirable to replenish the inventory immediately.

Their objective is to keep some of each item in stock at all times. Order point/order quantity techniques control each item independently of other items in the inventory. No attention is paid to the way components may be used together in combinations or in assemblies. Demand forecasts are usually based on historical demand data or statistical projections of such data using techniques like exponential smoothing. Forecast errors are measured and used to set safety stock levels. Some assumptions must be made about the pattern or "distribution" of these errors so that available mathematical tools can be used; this is really making the job fit the tools. In spite of all the built-in assumptions, calculating safety stocks is extremely powerful *in putting inventory where it does the most good.*

Such calculations can provide management with valuable information to help them set rational customer service policies. For example, Table 1 shows how safety stock would vary as service level

## Table 1 / Safety Stock on Bearing

Bearing BJ4726
Forecast = 2250/Week
Average Forecast Error = 450
Order Quantity = 9,000
Lead Time = 4 weeks

| Maximum Quantity Backorders per Year | Safety Stock Inventory | |
|---|---|---|
| | Pieces | Days Supply |
| 7,850 | 0 | 0 |
| 5,850 | 360 | 0.8 |
| 1,170 | 1,800 | 4 |
| 120 | 3,240 | 7 |
| 10 | 4,440 | 10 |
| 0 | 6,250* | 14* |

*Theoretically "infinite"

(measured in backorders) changed for a bearing with very stable demand. A good weekly forecast has held the error about equal to one-day's average usage. The safety stocks have been calculated using a measure of customer service defined as "percent of demand filled routinely from stock without backorders." If no safety stock were carried backorders would total almost a month's demand (7,850 pieces), probably occurring periodically in small amounts throughout the year. Required safety stocks increase faster as the tolerable level of backorders decreases. To reduce backorders by about 4,700 pieces requires approximately 1,500 more units in safety stock, but a further reduction of only 1,000 pieces in the backorder total requires an additional 1,400 pieces in safety stock. Theoretically it would require an infinite amount of inventory to insure "no backorders"; actually the maximum demands experienced on this item indicate that about 6,250 pieces would prevent backorders with only rare exceptions. Some companies, particularly those in the distribution or retail business like to measure customer service this way. Other companies are less interested in *how far* out of stock an item may go, and are more concerned about the frequency of its going out. They use a customer service measure defined as the percentage of times an item is "exposed" to a stockout that it does *not* stockout. An item ordered ten times a year will have ten "exposures" just before being replenished because its inventory is then at its lowest ebb and there is maximum danger of stocking out. A "90% service" requirement would aim for only one stockout per year, nine exposures out of ten without a stockout.

Many other measures of customer service may be of interest to particular companies but rigorous statistical calculations have been developed only for backorder and stockout measures. These methods give the appearance of great precision but the assumptions on which they are based make the results far from precise. If significant real benefits are to be obtained, these assumptions should be thoroughly understood, and the techniques applied with judgment and consideration of their *total* impact. Properly applied, however, they can be most effective in putting inventories where they will do the most good—the best service for the least investment—which is far better than rules-of-thumb like "one week's supply of safety stock on all items."

Table 2 illustrates this point for three items manufactured for

## Table 2 / Rule of Thumb  vs.  Statistics

| Item | Avg. Weekly Demand | Reorder Quantity | Rule of Thumb 1-Week S.S. | Rule of Thumb S.O. Per Yr. | Statistics Calc. S.S. | Statistics S.O. Per Yr. |
|------|------|------|------|------|------|------|
| P | 500 | 500 | 500 | 2 | 460 | 2 |
| Y | 500 | 500 | 500 | 9 | 965 | 2 |
| Z | 500 | 6500 | 500 | 0 | 0 | 2 |
|   |   |   | 1500 | 11 | 1425 | 6 |

stock. Carrying one week's supply as safety stock results in an inventory of 1,500 pieces and 11 stockouts per year. The table shows how statistical techniques could be applied to keep the safety stock total about the same but reduce the number of stockouts from 11 to 6. Also, stockouts are now evenly distributed among the items instead of being concentrated on one. *When safety stock is spread evenly, stockouts are not.*

## Material Requirements Planning

While there is a great surplus of published information available on order point/order quantity techniques, relatively little has been written about material requirements planning, although it is by far the older of the two techniques. Known as the Quarterly Ordering System, the technique has been used for a long time to plan procurement of materials for machine tools, shipbuilding, aircraft, locomotives and other heavy products. In it, a production schedule for building specific finished products in definite time periods was "exploded" (using the bill of materials of the finished product) into requirements for all sub-assemblies and components. New orders were scheduled to come in when existing inventories were used up. New order release dates were determined by subtracting the lead time to procure each item. The technique attempted to bring the items into stock or to the assembly area in time to match up with other components of the assembly. In the early days of many companies, when product lines were simple and sizeable backlogs of customer orders were on hand, this basic approach worked very well to plan replenishment orders.

As product models proliferated and as the product complexity increased, however, it became increasingly difficult to develop a practical, workable production schedule for finished products far enough in advance, explode all the bills of material, net out available stocks and trigger replenishment orders. The work of making the calculations and record comparisons was so time-consuming, the technique was impractical for products of even moderate complexity without use of a computer. Speed is essential because requirements for sub-assemblies and parts used in final assembly operations were determined first, while those for raw

materials and purchased parts were determined last, although they
are needed first. These difficulties led many companies to abandon
this technique in favor of the order point/order quantity system,
advertised as "a scientific technique to insure that all materials will
be available when needed." Hence, material requirements planning
was not applied in most companies that made many small or
medium-sized assembled products, chemicals, textiles, etc.

Because of the great amount of discussion in the literature, in
college courses, and by consultants on the sophistication of order
point/order quantity approaches, many companies have tried to
apply them to all items. Obviously, three of the basic premises of
order point/order quantity techniques are violated when making
assemblies. Usage of component parts is not uniform, as shown in
Figure 14, but is in lumps due to lot-sizing at higher levels; withdrawal of parts from stock is not in small increments because of the
lumpy usage. There is no need to replenish immediately after a
withdrawal if there are no requirements to make more of parent
items using the component.

The most significant problem which results from applying this
sound system incorrectly is the distortion of true priorities: knowing
what items are really needed in stock and when. The order
point/order quantity technique would trigger an order for the
component part in Figure 14, as shown in Figure 15, when the net
available inventory drops below the reorder point, and would indicate that the "due date" (when the item will be needed) is the end
of the planned lead time. For dependent demand items this is the
wrong date in practically every case. Figure 15 shows under MRP
when the component is really needed: when available stocks are
used up. Since demand is not uniform, components will be required
either earlier or later than the planned date using averages. The
formal order point has no way to update the priority to compensate
for changed conditions. The informal system of shortage lists and
expediting must take over to get those items needed earlier than the
planned dates. Those needed later are ignored, and account for the
large proportion of "Past Due" (but not needed) orders for both
purchased and manufactured items in most companies.

The greatest benefit to be obtained by using the right ordering
technique is proper priorities, so that attention can be focused on
getting the right items when they are really needed. The formal
system is then able to plan to meet true requirements, and there is

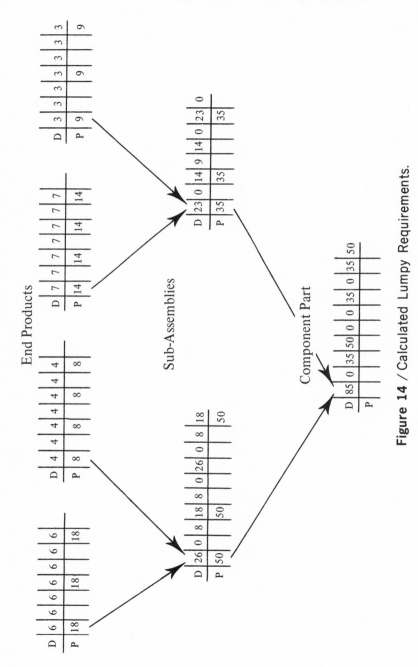

**Figure 14** / Calculated Lumpy Requirements.

Order Point

O.H. = 265    O.P. = 100    L.T. = 4 Weeks

| Week | 1 | 2 | 3 | 4 | 5 | 6 | 7 | 8 | 9 | 10 |
|---|---|---|---|---|---|---|---|---|---|---|
| Forecast | 20 | 20 | 20 | 20 | 20 | 20 | 20 | 20 | 20 | 20 |
| Actual Usage | 85 | 0 | 35 | 50 | | | | | | |
| Available | 180 | 180 | 145 | 95 | | | | | | |

Start Date — L.T. — Due Date

MRP

| Week | 1 | 2 | 3 | 4 | 5 | 6 | 7 | 8 | 9 | 10 |
|---|---|---|---|---|---|---|---|---|---|---|
| Required | 85 | 0 | 35 | 50 | 0 | 0 | 35 | 0 | 35 | 50 |
| Net Available | 180 | 180 | 145 | 95 | 95 | 95 | 60 | 60 | 25 | −25 |

Start Date — L.T. — Due Date

**Figure 15 / Priorities—Order Point vs. MRP.**

no need for informal systems. In far too many companies, the formal system is really used only to generate the paperwork to authorize those in the factory to take the action necessary to correct assembly line shortages, equipment downtime or idle manpower— *after the fact.*

Both OP/OQ and MRP ordering techniques are sound approaches, but until 1965 it was not really clear to practitioners when to use which technique. In that year Dr. Joseph A. Orlicky of IBM proposed the Independent/Dependent Demand Principle. "Independent demand" is defined as demand which cannot be related to that for other items manufactured by the company; typical of this is demand for finished products and spare parts sold directly to customers. For independent demand, the order point/order quantity system must be used, and it is necessary to forecast the demand expected. "Dependent demand" is demand for an item resulting from its use in higher-level components or sub-assemblies, and material requirements planning should be used. Only the end product or finished product needs to be forecast; requirements for all components can be calculated by "exploding" the scheduled quantity of the assembly through its bill of materials. This is one of the most important principles in the field today.

The underlying assumptions of material requirements planning are:

1/ A master production schedule can be developed for each finished model and projected far enough into the future to permit ordering raw materials and manufactured components soon enough to avoid shortages with minimum inventories. The planning horizon must extend far enough to cover the sum of the lead times required to procure a raw material or purchased component, manufacture it into a finished part, and make the several levels of sub-assemblies needed prior to final assembly. In many cases this can require a projection one year or more into the future.

2/ The product is well-defined in a bill of materials which shows all levels of sub-assemblies, finished or semi-finished components, raw materials and purchased parts required.

Material requirements planning and the order point/order quantity ordering technique are essentially opposites, as illustrated in Figure 16. MRP is an extremely powerful technique which is recognized as the best way to order components of assembled

| Order Point/Order Quantity | Material Requirements Planning |
|---|---|
| Part oriented | Based on product structure |
| Dependent on past | Looks only to future |
| Assumes averages | Handles erratic lumps |
| Gives "Start" dates— infers "Need" date | Shows "Need" dates— infers "Start" date |
| Aims to keep stock | Tries to run inventory down |
| Priorities inflexible | Keeps priorities up-to-date |

**Figure 16** / Comparison—Order Point   vs. MRP.

products. Here "assembled" should be interpreted to include components mixed, sewed, glued, packaged and chemically joined together, as well as those screwed, bolted and welded together. The economies of lot sizing and the protection of safety stocks can be attained with  material requirements  planning at least as well as with the order point/order quantity technique. Software packages to minimize the systems design and programming effort required to get MRP working for almost any company's products are available from most computer manufacturers and from many software houses and consulting firms. The design of such systems and use of these software packages will be discussed in greater detail in Chapter 8 on Developing Better Systems.

MRP has been called "The Missing Link" by those of us who have pioneered its application to modern business. It qualifies for this name in three ways. First, it makes it possible for *true "Need Dates" to be developed and kept up-to-date* with changes in real requirements. This is true for make-to-stock, make-to-order or both.

Second, it permits tracing which assembly cannot be made because one of its components is not available. We can now pinpoint *which products or customers will be affected if we cannot get all needed components made on time*. Now you can see who will get hurt, and when, if you oversell your capacity, don't solve manufacturing problems, or get hit with low yields or high scrap losses.

Finally, MRP forces a linking together of the forecast and overall inventory plan in the development of the master production schedule, and then provides the means to convert material requirements into meaningful data for capacity requirements planning.

If you combine component parts in any way in producing your products, you need MRP. You need it to determine when you really need materials, and to keep this information up-to-date, to decide when and how many to order to get maximum utilization of inventory, and to find out how much capacity you need and when. It will cost you more to be without it than you can afford.

## TOTAL INVENTORY MANAGEMENT

During an Executive Committee meeting in one company, the subject of inventory was being discussed. After hearing the usual comments—how the Controller thought turnover was too low, the Sales Manager wanted more emphasis on customer service, and the Manufacturing Vice President thought production costs and productivity deserved more attention—the General Manager in frustration asked, "How much inventory do we really need to run this business?" His question was answered by a loud silence.

Is there a rational answer? Most companies seem to follow cycles. When management's attention is focused on the high cost of money, the answer to how much inventory is needed is "less" (how much less depends on the amount of capital needed or on some arbitrary turnover target). When the need is to improve customer service by reducing backorders or adding warehouses, the answer is obviously "more" (hopefully not too much more). When costs come under close scrutiny, "more" seems necessary to take advantage of economical mass production manufacturing techniques (again, with little understanding of how much more). Inventory is viewed with biased perspective, too frequently *as a necessary evil*.

There's no doubt that the "right" total inventory will depend on management's goals for the key objectives of customer service and return on investment. What's missing in most companies, though, is a clear understanding of the relation between "return" and "investment" for inventories. Unfortunately, the standard, accepted cost-accounting methods of reporting inventory investment do nothing to clear up this misunderstanding. Without exception, they report total value of such classes as Raw Material, Purchased and Manufactured Components, Work-In-Process, Finished Product, Supplies and Miscellaneous. Why these classifications? Obviously they are convenient for "scorekeeping"; it's easy for accountants to track the flow of value from one to another. What comes in from outside sources is raw material or purchased components. When work begins, these are transferred to work-in-process, and labor and overheads are added. The result is finished product ready to ship to customers. The chain is completed by "accounts payable" and "accounts receivable." The supplies and miscellaneous classes are handy catchalls for items which cannot be identified with the product. Unhappily, these classifications contribute nothing that will help managers answer the basic questions involved in controlling inventory.

*How should inventory vary with changing sales?*
*What will happen to customer service and operating costs*
*if inventories are changed?*
*How much inventory is enough?*

I once saw a Controller's policy letter stating, "From now on, inventories will vary directly with sales rates. An increase of 10% in shipments will justify a 10% increase in inventory and vice versa." It's easy to be sympathetic to the frustration which initiated this. Inventory investment had been out of control for years, and was steadily increasing, usually more rapidly when sales fell. However, while it was perhaps better than no policy, his left much to be desired. First, he implicitly accepted the present total as right. Second, he viewed inventory too narrowly as simply supporting sales; he ignored its effect on operating costs.

The most popular measure of effectiveness of control of inventory is "turnover," as I stated before. It's usually expressed as the ratio of cost of goods sold in a period to the value of total inventory at the end of the period. Occasionally the inverse is used, expressing the inventory as a percentage of the total cost. Turnover has only one valid use—it is a *good measure of trends.* Its most serious deficiency is that it gives no clue as to what is "par" for your business.

Another common approach to setting inventory targets is based on "theoretical average inventory." When using order point/order quantity ordering, a time diagram of inventory looks like Figure 17, the familiar saw-tooth pattern. Theoretically, the average inventory will be one-half the order quantity, plus the safety stock for each item ordered.

Companies using this approach invariably find actual inventories considerably higher than the calculated targets because:

Usage is not uniform as the theory assumes, so the average is more than half the lot-size.

Parts are used in sets in assemblies, and shortages of a few components cause excess stocks of many.

Material is ordered earlier than planned to "be safe" or to "keep the shop busy." Few companies actually follow their ordering rules.

Many classes of inventory are not included: among them are work-in-process, build-ups of seasonal items, obsolete parts, and in-transit material.

Purchase discounts generate excess inventory above the calculated order quantities.

Even when efforts are made to account for all classes, the first three factors cause large differences between target and actual inventories.

Another common method bases the target level on time cycles. The flow of material for each major product family from raw material through finished product inventories is charted, with time cycles assigned to each phase. These are listed, and the total value of material calculated as shown in Table 3, starting with finished goods and taking into consideration the value added in each phase. The total of these phase values is then divided by the "relief" rate (the rate of sales or shipments) to determine a weighted-average day's supply as the target. It is usually assumed that "standard" time-cycles apply equally well to all items in the product family and to various "relief" rates. The results hardly justify the added effort,

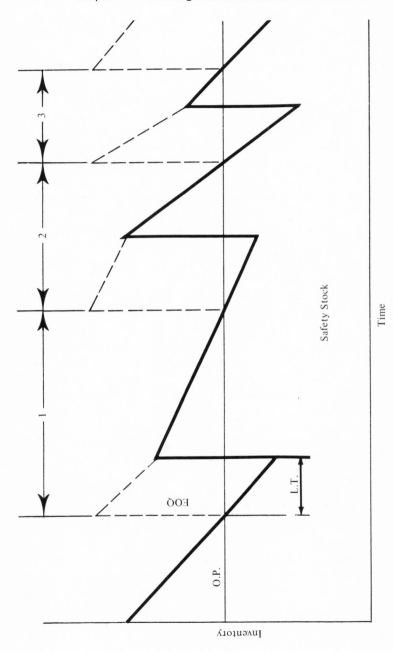

**Figure 17** / Time vs. Inventory—Order Point.

## Table 3 / Target Standard Inventory Calculation

Product Line I:

| Inventory Location | Planned Relief $/day | Standard Days | Standard $ |
|---|---|---|---|
| Finished Goods | 8,710 | 10 | 87,100 |
| Assembly W.I.P. | 8,710 | 1 | 8,710 |
| Compon. Stores | 8,260 | 10 | 82,600 |
| Sub-assem. W.I.P | 8,260 | 1 | 8,260 |
| Compon. Stores | 6,460 | 10 | 64,600 |
| Compon. W.I.P. | 3,500 | 30 | 105,000 |
| Raw Material | 1,700 | 30 | 51,000 |
| In Transit | 1,500 | 10 | 15,000 |

| | |
|---|---|
| Standard $ Total | 422,270 |
| Planned Finished Goods Relief | $8,710 |

Target Standard Inventory = 422,270/8,710 = 48-½ days

although the method does have the advantage of insuring that all phases of inventory are included.

Confusion and conflict will persist until we adopt a rational, professional view of inventories as assets, not liabilities. The present approach has failed because it obscures rather than clarifies the functions of inventories. They can and should earn a return on the investment, just like plant, machinery, tooling and other physical assets. The basic problems are to *determine how inventory earns a return and how much is earned.*

The conventional approaches, using accounting classifications, turnover rates, inventory percentages and the like cannot help solve these problems. Until they are solved, however, the relation of inventory to sales, its effect on costs and customer service, and the "right" total will remain mysteries. What's needed is *concentration on the functions inventory performs which yield measurable benefits.* These are summarized in Table 4 for six classes of inventory; the benefits associated with each are indicated.

Obviously, these functional classes bear no relation to standard accounting groupings. Inventories of raw materials, finished components, work-in-process, and finished products all could be carried to uncouple manufacturing operations, permitting them to run at the most economical rate and provide maximum utilization

## Table 4 / Inventory Functions and Benefits

| Function | Benefits |
|---|---|
| **Lot-size (or order quantity)** | |
| Uncouple manufacturing operations (i.e., screw-machines vs. assembly; supplier vs. user). | Purchased discounts; reduced setup, freight, material handling, paperwork and inspection expense, etc. |
| **Demand-fluctuation (or safety stock)** | |
| Insurance against unexpected demand (safety stock). | Increased sales; reduced outgoing freight, substitution of higher value product, customer service, clerical telephone & telegraph, packaging costs, etc. |
| **Supply-fluctuation** | |
| Insurance against interrupted supply (i.e., strikes, vendor lead time variations). | Reduced downtime and overtime, substitute materials, and incoming freight; increased sales. |
| **Anticipation** | |
| Level out production (i.e., to meet seasonal sales, marketing promotions). | Reduced overtime, subcontract, hiring, layoff, unemployment insurance, training, scrap and rework expense, etc. Less excess capacity in equipment needed. |
| **Transportation** | |
| Fill distribution pipeline (i.e., in transit, branch warehouse and consigned materials). | Increased sales; reduced freight, handling and packaging costs. |
| **Hedge (or speculative)** | |
| Provide hedge against price increases (i.e., copper, silver). | Lower material costs. |

of equipment. Equally obviously, the specific "return" from only a few of the minor benefits can be found in the accounting records of most companies; typical of these are freight, clerical expense and overtime. The most important benefits, however, like increased sales, lower setup costs and reduced downtime, lessening of hiring and layoff expense, have to be "engineered." The present accounting systems have been designed for the scorekeepers; they are of little use to the players making inventory decisions.

Solution of the problem of "how much inventory is enough" begins by evaluating the benefits for each functional class of inventory in Table 4. Theoretical approaches are available for evaluating return on inventory investment for each.

## LOT-SIZE INVENTORY (Finished parts in stock)

The "economic order quantity" was discussed earlier in this Chapter. Formulas balancing ordering costs with carrying costs to produce the lowest total have been apparently of more use to those teaching and writing about inventory than to those attempting to control it. Based on my surveys of those attempting to apply the formula in the real world, seven or eight have serious problems for every one able to identify specific benefits. The basic problems have been identified:

1/ Inaccurate data are used for projected demand (forecasts), ordering costs (estimates) and, particularly, the cost of carrying inventory (guesses).

2/ Attention is concentrated on individual items with the impact on total costs and investment ignored.

3/ Specific savings are not identified; the assumption is made that benefits will accrue automatically.

4/ Objective is limited to achieving the minimum total of ordering and carrying costs.

Most of those applying EOQ formulas have been playing a game of "magic numbers"—doing their best to get the "right" numbers to put in the correct formula and expecting it to make the proper lot-size decision, subject only to some modifications for rounding off or avoiding ridiculous extremes. The first to suggest a departure from this stereotyped approach was W. Evert Welch, who recognized that very significant savings could be made by applying the theory of EOQ *to "families" of items* even without evaluating the cost factors in the formula. Where intuitive or "rule-of-thumb" methods have been in use, Welch showed how to reduce the lot-size inventory of a family of items while continuing to place the same number of orders, or to reduce the number of orders while holding the lot-size inventory constant. Using his method, Table 5 illustrates the potential benefits of applying the theory of EOQ to a family of three items previously ordered four times a year. Welch pointed out that his results were not the "most economical," just *more economical than intuitive methods.*

In a similar approach, Dr. S.B. Smith showed how to force the EOQ formula to set lot-sizes for a family of items which would not exceed a restriction on the total capital available for this investment. He assumed that "economic orders" based on formula were already being used, but that the wrong carrying cost was employed. He showed how to calculate an "imputed" carrying cost which would achieve the desired percentage reduction in investment. Like Welch, Smith's concern was to get the maximum benefit from the theory for *groups of items subject to real limitations.*

About the same time several others had begun to question the validity of the search for the "correct" carrying charge. R.L. VanDeMark pointed out that changing the value assigned to the cost of carrying inventory would be a logical way to vary the total lot-size inventory as desired. R.G. Brown called this factor "a management policy variable," and also discussed using the carrying charge as a device to control total lot-size inventory and the expense associated with equipment setup.

In 1963 we developed a technique called LIMIT (*L*ot-Size *I*nventory *M*anagement *I*nterpolation *T*echnique), by which the relationship between investment in inventory and total ordering costs for large families of items could be presented in tabular form or as a "trade-off" curve, as in Figure 18.

Intuitive or rule-of-thumb ordering result in inventory/cost relationships above the theoretical curve, like point "X" on Figure 18. *The theory can then be used first to get down on the curve.* Several alternatives are possible: hold costs the same and reduce inventory (move vertically); hold inventory constant and cut costs (move horizontally); or cut both inventory and costs (move

**Table 5 / Optimum Inventory for Limited Orders**

| Item | Annual Uses $ | $\sqrt{A}$ | Present Lot Size | Present Orders Per Year (N) | New Lot Size | New Orders Per Year |
|------|---------------|------------|------------------|----------------------------|--------------|---------------------|
| A | 10,000 | 100 | 2500 | 4 | 1100 | 9 |
| B | 400 | 20 | 100 | 4 | 220 | 2 |
| C | 144 | 12 | 36 | 4 | 132 | 1 |
| Total | 10,544 | 132 | 2636 | 12 | 1452 | 12 |
| Avg. Inv | | | 1318 | | 726 | |

$$K = \frac{\Sigma\sqrt{A}}{\Sigma N} = \frac{132}{12} = 11 \quad \text{New Lot Size} = K\sqrt{A}$$

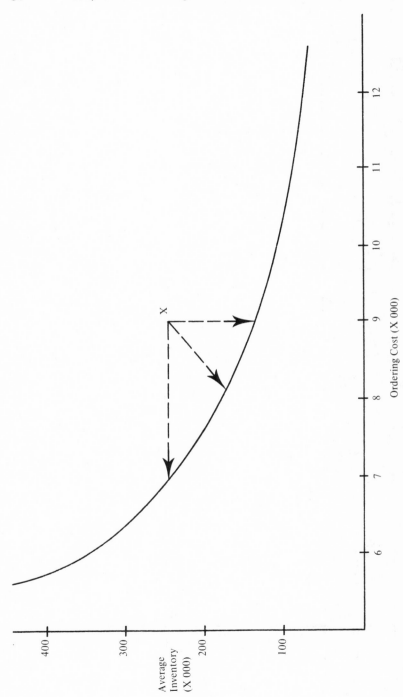

**Figure 18** / Lot-Size Inventory vs. Ordering Cost.

diagonally). The real value of the technique is that it provides a management tool to determine when greater savings in ordering costs represent an inadequate return on the investment in additional inventory. *It thus provides a rational way to set total lot-size inventory goals.*

Achieving these goals, however, will not be "automatic." If costs are to be reduced, *specific actions must be identified and taken—* setup men must be given additional work, clerks must be transferred, etc. The techniques show the potential savings; management must act to capture them. Even where the goal is reduced inventory, the *formulas* alone *will not generate savings.* Tompkins has shown how just changing order quantities simply shifts total inventory between lot-size and safety stock *unless capacity is adjusted.*

Companies have used aggregate techniques in a variety of ways to achieve significant benefits in operation. Individual item order quantities can be set to insure a full work load and minimum total inventory for the actual number of setup men or order clerks working. They can also be calculated to insure that the total setup time on a critical machine (taking it out of production) does not cut into the minimum operating time needed to produce the required total output. In this situation, the techniques can also provide additional data to aid capital investment decisions by showing the potential reductions in inventory from adding equipment, thus removing setup limitations and permitting smaller lot-sizes to be run.

Purchase discount decisions have suffered from the same "one-item-at-a-time" approach as used to set order quantities. The decision to take the discount or not is usually based on whether the savings from lower unit prices and fewer orders exceeds the cost of carrying the additional inventory. The decision hinges on "knowing" the carrying cost, and the approach gives no indication of the total impact on capital investment of *all* discount decisions which might be taken. Viewing the *net* savings as return on investment and ranking a number of potential discounts in order of return on the added investment would permit taking only those discounts which meet management's desired return, similar to goals set for capital equipment. The objection is frequently raised that "we can't identify now all items we may be able to get discounts on." This is true, of course, but no excuse for not using the technique for all known discount situations, and making a real effort to be sure all "A" items (those with high usage-value) are covered. These will obviously generate the bulk of the return, and are most likely to have discount possibilities.

## LOT-SIZE INVENTORY( Work-in-process)

With minor exceptions, the bulk of work-in-process should be lot-size inventory. For castings being machined into finished parts, Figure 19 shows the lot quantity, "q", expressed in "pieces" and the lead time in "weeks" needed for all operations on each lot. Averaged over the year, this work-in-process will be q(LT/52) for each lot. The total number of lots, "n," run during the year will result in an average total work-in-process of nq(LT/52) for each item. Since "nq" is the total annual demand for the item, the *average* amount of pieces of work-in-process for each item will then be:

$$WIPavg=Annual\ Demand\ (LT/52)$$

The interesting point is that the *average work-in-process is not affected by the lot size;* it is proportional only to the annual demand and the lead time. This makes it simple to set target levels for work-in-process if "normal" lead times are defined. It also demonstrates clearly the advantage of cutting lead times and reducing work-in-process in the plant, particularly for high-usage items.

Working with dollars and real inventories, scrap, rework, material costs, labor and overhead must be introduced. Increasing lot-sizes to make provision for scrap losses will increase the average value of work-in-process. Rework will also increase it by the longer time and additional labor required. The average *value* of a lot depends on how material and labor are added as the lot is processed. When the bulk of the labor is performed in the first few operations, as shown in Figure 20, the average value will be closer to the finished total. When the bulk of the labor is added in the finishing operations, the average value of work-in-process is close to the material value. For most manufactured parts, all material value is committed when the lot is started. Subassemblies and assemblies may have some components added in later assembly operations; this would materially affect the average value of work-in-process

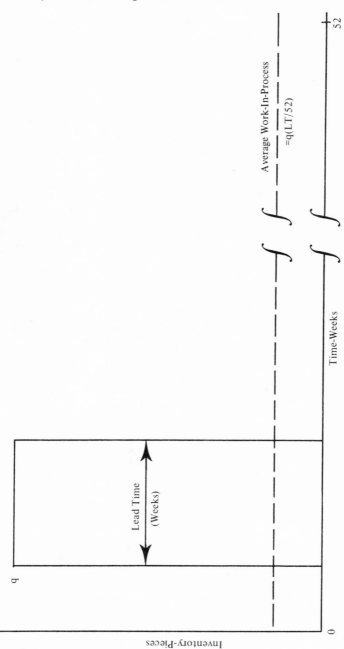

**Figure 19 / Average Value of Work-In-Process (#1).**

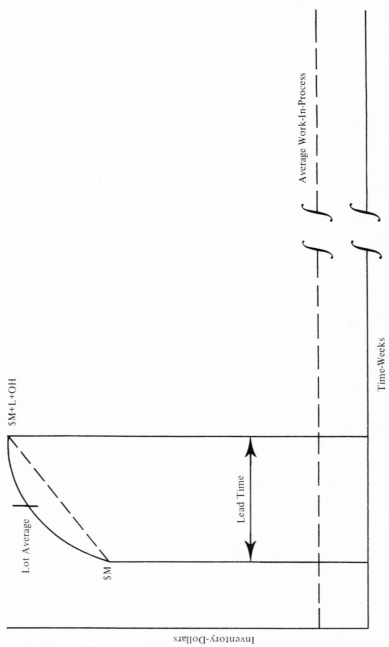

**Figure 20 / Average Value of Work-In-Process (#2).**

only for long assembly cycles, however, where all parts for the assembly were not drawn from stores and issued together.

## DISCRETE LOT-SIZING TECHNIQUES

Studies are underway to evaluate the error involved in using the square-root calculation when planning total lot-size inventories for families of items, although discrete techniques are used in the actual ordering. It is already clear that this error is small compared to the benefits achieved. Since one-year planning horizons are typical of computerized material requirements plans, computer programs can be developed to get better answers by simulating aggregate lot-size inventory investment vs. ordering expenses based on the specific discrete lot-sizing techniques used.

## DEMAND-FLUCTUATION INVENTORIES

Usually called safety stocks, demand-fluctuation inventories were discussed earlier in this chapter. Using statistical methods for determination of safety stocks, a "trade-off curve," Figure 21, can be drawn for families of items. This shows the relationship between inventory and customer service, and has the same general shape as the lot-size inventory curve of Figure 18. The amount of inventory increases dramatically as the customer service level approaches perfection. This is also shown by the data in Table 1.

Rules-of-thumb, intuitive approaches or other non-rational methods of establishing safety stocks usually result in "off the curve" performance, point "X," that is, too much inventory is carried for the service level actually achieved. Application of the theory permits getting down on the curve, either by holding service level constant and reducing inventory, or holding inventory constant and improving service or some combination of both. The technique also gives data needed to evaluate the "return on investment" implications of moving from one point on the curve to another by changing policies. Return is expressed in terms of improved customer service and not in dollars. Improved customer service should generate some benefits in increased sales, fewer substitutions of higher cost products, or reduced expenses in

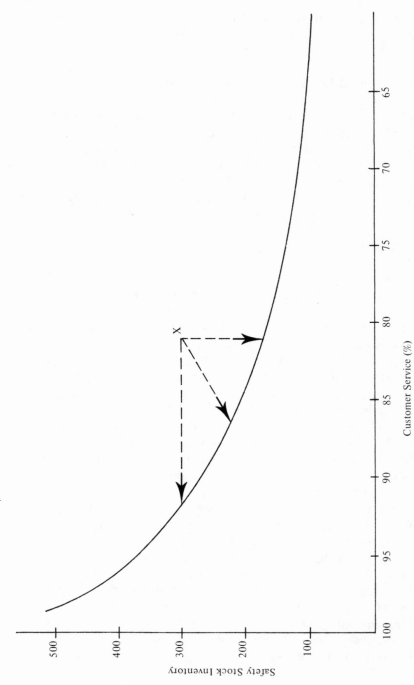

**Figure 21 / Customer Service vs. Safety Stock.**

freight, handling backorders, answering customers complaints, expediting, etc. The most important payback will usually be increased sales, and the proper customer service level must be determined by a manager's judgment based upon estimates of how sales might be improved. *There is no way to calculate which is the "right" point to be on the curve.* However, the trade-off curve technique does provide quantitative data for decision making, thus putting into perspective the investment implications of policies like, "We can't afford to ever be out of stock" and other beautiful generalities.

## SUPPLY-FLUCTUATION INVENTORIES

Uncertainty of supply affects both independent and dependent demand items. Variations in replenishment lead time probably cause more stockouts, backorders, and missed delivery dates than fluctuations in demand. Theoretical approaches to combining the effects of both demand and lead time variations in calculating order points have been developed but are rarely used. Raising order points on manufactured parts to provide more safety stock as lead times get longer aggravates the basic cause of the longer lead times—higher levels of work-in-process. True control of lead time requires regulating work input rates to match available capacity and provide a balanced flow of work through manufacturing facilities. More on this will be discussed in Chapter 6.

This basic input/output principle applies equally well to purchased materials. Falldowns in vendor deliveries are caused more frequently than is recognized by the customers' poor ordering practices. "Dynamic order points" would react to such delivery failures, trigger more new orders and make the problem worse. The professional inventory manager recognizes that changing safety stock must be handled carefully, and not left to an automated mathematical formula.

Unlike the order point/order quantity system, theoretical approaches to setting safety stocks for dependent demand items using material requirements planning have not yet been developed. Demand-fluctuation inventories can be carried either in finished product or in additional sets of parts not actually assembled. The latter act effectively as safety stocks, so it would appear to be unnecessary to carry supply-fluctuation stocks on dependent demand items.

There are two common exceptions to this. Most companies feel that purchased raw materials and components are not under their control and prefer to carry stock to guard against failure of a vendor to deliver a new lot on time. A safety stock equal to one extra lot would appear to be the maximum required; the minimum might be the amount usually withdrawn from inventory for the largest machining or assembly lot of the next higher level component. In many companies, semi-finished components or assemblies can be carried in stock and made into the finished product in a very short time. There may be relatively few varieties of semi-finished components assembled into a large number of finished products, so the most flexible inventories are the components. Mechanics' and carpenters' hand tools held in bulk inventories, but sold in mixed sets, are a good example of this situation. Carrying safety stocks at this level results in very much lower total fluctuation inventories, since the total demand variation is likely to be much smaller than the sum of the demand variations of the individual finished items. Attempts have been made to apply rigorous statistical techniques to determining the needed amounts of fluctuation inventory for such items, so far without much success. Carrying safety stocks only at the high and low levels implies good control over processing lead times, which requires sound capacity planning, good work-input control, and effective shop floor control, with a dynamic priority technique to keep the system responsive to the inevitable changes.

Within a manufacturing plant there are several basic causes of failure of supply, each requiring its own method of evaluating the needed amount of fluctuation inventories to give a desired level of protection.

*Machine breakdowns, equipment malfunctions or tool failures* require inventories to prevent down-time. Manufacturing people know that some work-in-process is needed to keep people and machines busy when operations that feed them stop. Most plants count on high levels of work-in-process to provide this kind of insurance automatically, but it doesn't. Just having plenty doesn't provide the right items. If machines break down frequently, tooling is cranky, or people can't consistently produce good quality, work-in-process inventory should be *planned* to keep subsequent operations busy. Pick active items, with plenty of work-content for the operations to be fed. Hold one lot or enough to make up for the interruption. The investment in this insurance can be evaluated; if it's high enough it can help focus attention on solving the real problems. Again, it's a question of having return on investment data to help make better decisions. While such calculations may be tedious, they are not difficult. The problem is deciding how much it will cost in improved preventive maintenance programs and better designed tooling to prevent the breakdowns and make the inventory unnecessary.

*Quality rejections* can result in small scrap losses or even unexpected loss of whole lots, such as in pharmaceutical production, food processing, heat treating, ceramic firing and the like. Small scrap losses are usually covered by increasing the starting lot-size so that the good pieces meet the requirements. Loss of the whole lot, however, requires specific planning to carry enough finished product to last until a replacement lot is made, or carrying a safety stock in semi-finished products or components already through the critical operation to insure a continuing supply. Again, the computations of the needed total inventory are tedious but not difficult, and the problem is one of determining the potential cost of curing the problems which cause the periodic rejections. It should not be overlooked that replenishment of rejected lots actually requires increased total capacity to handle the thru-put. Replacing rejected material wastes capacity as well as causing material and labor losses.

*Temporary recurring bottlenecks* result from failure to balance material flow through manufacturing facilities. Several jobs arrive simultaneously at a work center and compete for available capacity. This occurs most frequently at service departments like painting, plating, heat treating and inspection. While no "scientific" approach is known, a good practical approach is simply to count daily the backlogs ahead of these departments over a period of time. The number of jobs in queue will be found to vary from a lower to an upper limit, say from 5 to 15. Obviously the facility *never uses and doesn't require the minimum quantity of 5 jobs* in the queue, since it is always there. Work-in-process levels could be reduced by 5 jobs with danger of only occasional downtime. The queue range would be 0-10 and the average level 5 jobs *even if control of flow weren't improved.* Such determinations can be made easily for critical areas where large bottlenecks occur frequently.

*Strikes or threats of strikes* at vendors' plants, by truckers, or on railroads require planned increases in inventories to insure supply during the strike. Potential strikes within a plant may cause similar inventory buildups. Most trade journals carry information on important labor contract expiration dates so that the timing of buildups can be determined early in the year. Such inventories usually represent a very substantial investment. Deciding to carry them assumes it is more expensive to shut the operation down, let customers wait or work on alternate materials in the interim. Determination of the investment in fluctuation inventory is simple. The return on this investment is rarely considered. Because of the impact not only on a company's economic health but on our overall economy, such decisions should be much more carefully considered than they usually are.

## ANTICIPATION INVENTORY

Peaks and valleys in sales demand are caused by seasonal products, sales promotions, and selling price changes. There are significant savings to be made by carrying anticipation inventory to level out production in spite of such swings. Anticipation inventory is needed also to cover interruptions in production from relocating or replacing machinery, vacation shutdowns or moving departments or plants. The savings are in two classes: lower investment in plant and equipment compared to having capacity to meet peak loads, and reduced costs such as hiring, layoff, training, overtime, downtime, scrap and rework. As usual, these costs are not detailed in standard accounts but must be "engineered." The inventory required will be the total of all items needed in excess of the capacity of the manufacturing facility.

Techniques for planning capacity to meet peak demands or interruptions in production are well known. These will be discussed in more detail in Chapter 6. Table 6 is typical of plans for a make-to-stock business. It shows production rates required to meet forecast shipments, and build up the finished goods inventory in anticipation of the annual vacation shutdown in July. The intent is to have total finished goods inventories at the end of the vacation

## Table 6 / Production Plan

| Week | | Sales | | Production | | Inventory |
|------|------|------|------|------|------|------|
| | | Week | Cumul. | Week | Cumul. | Total |
| Start | | | 42,000 | | 38,500 | 15,000 |
| 4/7 | Plan Act | 3,000 | 45,000 | 4,000 | 42,500 | 16,000 |
| 4/14 | Plan Act | 3,000 | 48,000 | 4,000 | 46,500 | 17,000 |
| 4/21 | Plan Act | 3,000 | 51,000 | 4,000 | 50,500 | 18,000 |
| 6/23 | Plan Act | 3,000 | 78,000 | 4,000 | 86,500 | 27,000 |
| 6/30 | Plan | 3,000 | 81,000 | 4,000 | 90,500 | 28,000 |

shutdown no lower than the base level (equal to the target lot-size plus demand-fluctuation inventories). At the start of the vacation period, therefore, the inventory would have to be above base level by an amount equal to total shipments during the shutdown. To build up this figure requires that production be higher than the shipping rate. The amount of the difference, of course, depends on the time available to build up the inventory. The farther in advance this buildup can be started, the less the difference between average and actual production requirements. Ideally, such inventory buildup should start at the close of the preceding vacation. The inventory at the start of the planning period ($I_s$) plus production (P) during the period minus total shipments (S) will be the inventory at the end of the planning period ($I_e$). The calculations are very simple:

$$I_s + P - S = I_e$$

This is a basic relationship for total inventories. It shows clearly that *inventories are really controlled only by balancing production* (*input*) *and shipments* (*output*). The equation can be used to determine any one variable when the other three are known. For vacation buildups, forecast shipments, the starting and ending inventories would be known, and the calculated production can be spread over the planning period. If production is inadequate to meet shipments, the formula can be used to determine what total inventories will be at the end of the planning period. This can then be compared to target levels for lot-size and fluctuation inventories to determine how customer service and/or operating costs might be affected. The formula can also be used to determine the relationships among inventory, input and output of any manufacturing facility: work centers, departments, plants, etc.

Capacity plans like Table 6 permit setting weekly or monthly total inventory budgets which provide the basis for control of capacity. Comparing actual to planned figures *for all three variables (shipments, production and inventory) on a single report* provides all the facts relevant to decisions about whether or not to change production rates. Capacity plans are also an excellent device for comparing alternative strategies for production and inventories. For example, if seasonal shipments are met by level production throughout the year, maximum swings will occur in the inventory. Peak inventory can be minimized by changing the production rate one or more times during the year; the alternative production rates and inventories can be studied to evaluate costs and benefits.

Inventory investment is easily determined but costs are more difficult. In addition to hiring, layoff and the other  factors mentioned, the intangibles of loss of skilled help, reputation as a good place to work, and the social responsibilities of business to provide dependable employment must be considered in making production decisions. No formula or technique will make these decisions for management.

For make-to-order businesses, the backlog of orders corresponds to product inventories in make-to-stock companies and a similar equation can be set up:

$$B_s + 0 - P = B_e$$

The order backlog at the start of the planning period ($B_s$) plus incoming orders (0) minus shipments (which are equal to production (P) since there is no finished-product inventory) will be the resulting backlog at the end of the planning period ($B_e$). Production plans using these relationships can be set up like Table 6, and used for the same purposes of  control and decision-making.

The major problems in planning anticipation inventories are the difficulty in forecasting seasonal peaks, and the effect of sales promotions, selling price changes, shipments, and incoming orders a year in the future. To be most effective, therefore, such plans must be updated regularly as forecasts change or production schedules are missed. Another problem arises in evaluating actual inventory in fluctuation and anticipation groupings. It is possible to determine the actual value of lot-size inventory on hand. However, total inventory above this amount is combined fluctuation and anticipation classes, and there is no way to distinguish between them. All that can be done is to compare the actual total of all classes against the plan.

## TRANSPORTATION INVENTORY

This classification includes all materials in the market distribution pipeline in transit, and therefore not available to serve the other functions of inventories. There is some minimum in-transit inventory necessary just to keep the operations going. The return on this investment is the value of not shutting down. Calculating how much is a straightforward determination of the number and value of shipments in transit. If the shipping time is two weeks, and a truckload worth $40,000 is shipped each week, the in-transit inventory will obviously be $80,000. For a given sales rate, the only way to reduce this inventory is to reduce the transit time. This usually involves higher freight costs. "Not spending" this money can be viewed as a "return" on the amount of transportation inventory investment above that for the faster shipments.

Rather dramatic reductions in air freight costs in recent years (from 27¢ per ton mile for DC3 flights to an estimated 3 to 5¢ per ton mile for 747 jumbo jets) may make significant decreases in in-transit inventories economical for companies with high-value, low-mass products. Although it is still considerably more expensive than truck, rail and boat rates, air freight makes possible lower inventories from reduced in-transit and smaller safety stocks; the "return" generated by using the cheaper shipping methods could be too little payback on the extra inventories necessary. An additional factor is the potential reduction in accounts receivable. One machine tool company flew a large piece of equipment worth

$280,000 across the ocean. Bringing the cash into their operations several weeks earlier was a major factor in their decision.

Planning and controlling all classes of inventories in branch warehouses should be based on determining the economic frequency of shipment. This aims for the minimum total of ordering, shipping and inventory carrying costs. The total lot-size inventory will be set by the shipping frequency and standard statistical analyses used for setting total safety stocks. Adding warehouses will require more in-transit material, and also more fluctuation inventories to maintain the same customer service levels.

## HEDGE INVENTORIES

Companies using large quantities of basic minerals like silver and copper, or commodities like wool, grains, etc., can make significant savings by buying extra quantities when prices are low. Also, extra quantities of purchased items brought into inventory in advance of a price increase will reduce the total cost of these items. These savings are obviously the return on the added inventory investment. For both these examples, the quantity to be purchased should be determined by the potential return on the investment, modified of course by such factors as available capital, product shelf-life, and potential design obsolescence.

## ANSWERING THE BASIC QUESTIONS

Too few companies are enjoying real benefits from "scientific" inventory planning techniques. Understanding proper use of the trade-off analyses presented here, a few professional practitioners are able to give quantitative, rational answers to the basic question of inventory management.

### *"How should inventory vary with sales?"*

The answer is obvious when the functional classifications are studied individually. Lot-size inventory should increase or decrease approximately *as the square-root* of sales. Doubling sales requires only forty percent more inventory to keep total costs minimized. This is self-evident when the familiar square-root formula is used for setting order quantities. It seems likely that discrete lot-sizing techniques produce similar results, but this is yet to be proved. The work-in-process fraction of lot-size inventory has already been shown to vary directly with annual sales data.

The way demand-fluctuation inventories change with sales rates depends on how forecast accuracy is affected. If increased sales result from selling more of the same things, demand variations usually decrease, meaning that forecasts are closer to actual demand. Safety stocks to maintain a given customer service level will not increase proportionately to sales; in many cases they can be held at the same level, and sometimes even reduced. When higher sales are generated by adding entirely new products or more models of existing lines, demand-fluctuation inventories will frequently have to be raised well above proportionate levels. New products are the most difficult to forecast and, hence, errors are large.

Supply-fluctuation inventories apparently bear no rational relationship to sales rates. A good practical approach would seem to be to hold target levels the same for small ($\pm 20\%$) changes in sales rates, and change them only for larger swings. Since few companies experience growth-rates or cut backs in excess of 20%, this factor can often be neglected. Anticipation, transportation, and hedge inventories will all have to vary directly and proportionately to sales rates, since all involve "time-periods of supply."

An awareness of these relationships makes it possible for the professional to adjust budgeted inventories for varying forecasts of sales demand on a rational basis *working with the functional classes.* Translating into the traditional accounting groupings requires tedious but not difficult calculations.

> *"What will happen to customer service and*
> *to operating costs when inventory levels*
> *are changed?"*

This can be answered through the development of trade-off curves like Figures 18 and 21. The Stanley Works, Fafnir Bearing, McGill Manufacturing, Black & Decker and other companies have used the aggregate lot-size approaches discussed here to develop inventory investment vs. operating cost data. Order quantities will then be consistent with real-life limitations on setup or equipment capacity. New dimensions can also be added to capital equipment justification analyses when inventory investment changes can be evaluated.

A very simple approach, used by many companies for deciding how low work-in-process can be dropped before down-time costs are affected, is to watch the "queues" of work in major work centers, and keep reducing these backlogs until they occasionally drop near zero. This generates many fringe benefits beyond reduced inventory. It focuses attention on smoothing out the flow of work so the plant can run with lower work-in-process and fewer interruptions. It produces shorter and more dependable lead times which lead directly to improved customer service. It's well worth the extra management effort required.

Fewer companies have used customer service vs. investment trade-off curves, but those who did gained considerable insight into sales policy implications, and saw real improvement in working relations between marketing and manufacturing people. Probably the greatest benefits have come to those who used the approach to improve *both* service and investment. You can't do better than to make both the controller and sales manager happier.

Probably the most interesting application of the total approach was developed at Fiber Industries, Inc., Salisbury, North Carolina. This plant manufactures synthetic staple fibers to stock, and incurs very heavy changeover costs on equipment, which usually runs near full capacity. Applying both statistical safety stock and economic lot-sizing techniques, their computer program provides data to permit them to draw curves like Figure 22, showing inventory vs. changeover cost for a range of customer service levels. With these data, updated weekly, management is able to study a wide variety of alternatives.

For example, if total inventory is to be kept at the present level ("X"), the curves show how much operating costs will increase to achieve a given improvement in service (move to "Y"). They also show how service will be affected if inventory is to be reduced and operating costs held constant (move to "Z"). Management can see the alternatives of the various courses of action available and the trade-offs involved.

In one company the traffic manager investigated alternate shipping costs from the East to the West coasts, and estimated that annual freight expense could be reduced about $50,000 using boat shipments via the Panama Canal instead of direct rail. The

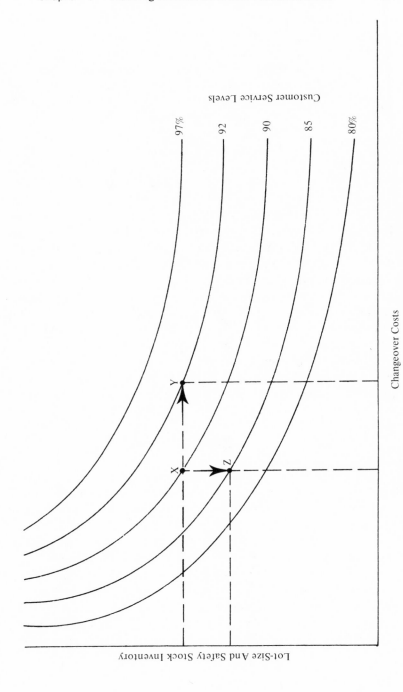

**Figure 22 / Inventory vs. Changeover Costs and Service.**

production control manager pointed out that the increased transit time would result in seven times as much inventory in transit. In addition, longer lead times would require almost four times as much safety stock in the West coast warehouse to maintain the same level of customer service. The $50,000 annual freight savings represented an inadequate return on the added capital investment in inventories, and the company continued shipping by rail.

*"How much inventory is enough?"*

There is no straightforward, fixed answer to this. *It is really dependent on management's policies.* What change do they want in customer service? What increases in costs will they accept to reduce the total investment in inventories? What are their policies regarding market distribution and warehousing? These and other questions relating to inventories will determine how much inventory is required. Equally important are alternate uses of capital and the return they can generate. The return that could be generated if capital were invested in another plant, new equipment, or research and development projects instead of in inventory must be taken into consideration to answer this deceptively simple question. The professional approach described here can aid in setting policies by making clear the trade-offs involved. Inventory need not continue to be the unmanageable mystery that it now is in most companies.

## CONCLUSION

Inventories must be viewed as capital assets, not unavoidable liabilities. There are techniques available which provide practical, usable tools to study the return on investment in inventory. While not precise, such quantitative data can help eliminate useless discussion of vague, general objectives like "better customer service," "higher turn-over," and "less upset in the plant."

Inventory total values expressed in the traditional accounting classes contribute nothing to determining return on investment. It is clear that new accounting methods are needed to provide data on both actual investment and related costs in the form most useful to operating managers. The development of these methods is a real challenge to the accounting profession to furnish management with information for true control as well as for good scorekeeping.

The *functions of inventory which generate savings* must be studied to see how available techniques can be applied to produce tangible benefits in lower costs, smaller inventories or both. Without changing total inventories, theoretical techniques can be used to put inventory where it does the most good. Contrasted to intuitive rule-of-thumb approaches, application of the theory permits "getting down on the optimum curve" and generating the highest possible return on inventory investment at that level.

These techniques can then be applied to determine the return on investment implications of moving from one point on the trade-off curve to another. This permits quantifying the relationships between increased inventories and reduced costs or improved customer service to give real facts in familiar terms to managers making operating decisions. The alternate, overworked approach of trying to find the "right" formulas and "magic numbers" to put in them *to make the decision for the manager* has not worked and will not. The effects on the strength of a company are too important to trust inventory decisions to machines using automatic rules. Systems must be developed to provide men called managers with data for decision-making.

# 6

# Being Sure You're Making Enough Product

The question, "Are we making enough in total?" is at least as important as the four basic questions of inventory planning discussed in the previous chapter. Capacity planning is one of the basic elements of an effective manufacturing control system. Unless it is soundly conceived and executed, the rest of the control system cannot be effective; inventory planning techniques will not function properly, the plant will fall behind schedule, shortages will increase, and customer deliveries will be missed. One of the very few truisms in the field is, "If you aren't making enough in total, you will not be able to make the right things." *You need both priority planning and control and capacity planning and control.*

The need for capacity planning is widely recognized. Practically every company makes annually an estimate of plant and equipment requirements based upon forecasts of one, two or more years' product demand. Direct manload needs are also calculated, and expense budgets developed for the planned level of operations. The basic question which is answered is, "What do we need in plant, equipment and manpower to be sure we are making enough product in total?'' Unfortunately, in too many companies, this planning is not updated as the year progresses. Day-to-day decisions regarding manpower, shift hours, and overtime are made by expediency rather than being based upon a sound plan. *It is not enough to answer this question once a year; it must be reevaluated constantly.*

All inventory planning techniques depend upon the validity of planned lead times. This is true of both purchased and manufactured items. The order point/order quantity approaches use the planned lead time to calculate both the expected average demand and the amount of safety stock to be added to it. Material requirements planning uses this planned lead time to develop the dates on which work should be started in order to have components in time to meet projected end-product requirements.

The most common opinion seems to be that lead times are ''ordained,'' meaning determined by factors external to ordering systems, and that the systems must be ''tuned'' by adjusting lead times when necessary. Most of those managing ordering systems seem to feel that they are ''victims'' of varying lead times and have no control over them. Neither of these premises is true. Control of lead time begins by developing a sound capacity plan for each manufacturing facility.

A manufacturing plant in operation is very much like a funnel. Its output capacity, while not fixed, is certainly difficult to change quickly by any significant amount. The work input rate is determined by how many customers order how much product and whether it is made to stock or to order. When this input rate exceeds plant capacity, the funnel fills up—backlogs of unstarted orders or work-in-process or both increase. When the funnel is too large and customer orders are not adequate to generate enough input, work-in-process and backlogs will dry up unless some artificial demand is created by ''make work'' orders released to keep the plant running. Of course, any real factory is really a series of funnels, one manufacturing facility feeding another, and the problem is to keep all the funnel outputs in balance so that overflows and dry spots are avoided.

## CAPACITY MUST BE ADEQUATE

Lead times for manufactured or purchased items can be controlled, but only when capacity is adequate. This is the first of three requirements. When order backlogs or work-in-process levels change because of the failure of capacity to match demand, lead times must also change. As backlogs and work-in-process levels get higher, delays caused by "waiting in line" get longer, and the *average actual lead times* of all work going through a manufacturing facility will get longer. It is obvious that backlogs determine average lead times.

What is not obvious is that as average lead times get longer, specific actual lead times on individual items can get more variable, less predictable. A good rule of thumb is that the actual lead time of any item can vary from the planned average lead time by one less than the average. With an average lead time of fifteen weeks, the actual lead time for any item may be fifteen weeks *plus or minus fourteen.* This is equally true of manufactured and purchased items. Under extreme pressure you can get most badly needed materials in a week; on the other hand there is always a very significant number of "Past Due" open orders not delivered within the planned lead times—and mostly not needed yet.

Inadequate capacity has the dual effect of increasing average actual lead times and of making specific actual lead times less predictable. Both make inventory planning less reliable, and it's hard to say which is worse. Excess capacity, on the other hand, can only result in either downtime (and low productivity) or excess inventory. Since manufacturing people detest downtime more than kids hate haircuts, too much inventory is inevitable. You can't win unless capacity levels are right.

## INPUT SHOULD BE SMOOTHED WHEREVER POSSIBLE

If followed rigidly, both order point/order quantity and material requirements planning ordering systems generate a very erratic rate of release of work. One week's orders released to be started may be a fraction of or many times that of the next week. Figure 23 shows how a soundly-designed material ordering system would like to release work to one company's facilities. While the average is 2700 standard hours, the minimum week is one-eighth of the maximum. To avoid running out of work in week 24, the foreman of this department would have to maintain a backlog of *at least 4 weeks' work*. This extra lead time could be eliminated if level release of work is imposed on the ordering system by input control. Where this can be done as a separate activity, significant amounts of lead

| Week | | Week | |
|------|------|------|------|
| 14 | 4500 | 19 | 1200 |
| 15 | 3900 | 20 | 700 |
| 16 | 4800 | 21 | 900 |
| 17 | 3800 | 22 | 2000 |
| 18 | 4600 | 23 | 600 |

**Figure 23 /** Standard Hours Released Work.

time can be eliminated from the inventory planning system, and its ability to give valid priorities improved substantially. Exactly the same is true of purchased materials obtained from outside manufacturing sources, particularly when the customer "buys" a large share of a vendor's capacity. There is nothing inherent in material ordering techniques to cause them to match work input to capacity. Erratic, uncontrolled flow of work through the plant produces the same effect on non-starting operations. These are much more difficult to control, however.

If the material planning system does nothing to insure against operations running out of work, the manufacturing people will. They'll maintain substantial backlogs of work so they never run out in lean input weeks. These backlogs—on paper or in work-in-process—cause long average lead times and widely variable actual lead times on specific items. *Erratic work flow results in variable, unpredictable lead times.*

## LEAD TIMES MUST BE DEFLATED

It is now well-recognized that we can't "make on-time delivery more certain" or "make it easy for the plant to stay on schedule" simply by allowing more lead time to get work done. Increasing planned average lead times just triggers more orders to be released, dumping more work into your plant or your vendors', increasing queues, lengthening actual average lead times, and making actual lead times on specific items more erratic.

*Instead of helping, inflating lead times makes all the problems worse.* The temptation to extinguish this fire with gasoline is almost irresistible, however. Until you recognize that lead time is both a cause and an effect of backlogs, it's hard to understand that lead times are like mini-skirts—the shorter the better.

## GROUNDWORK FOR CAPACITY PLANNING

Before making a capacity plan there are several questions to be answered which require management decisions. They are:

1/ What forecasts will be used for demand? How frequently will they be reviewed?
2/ What changes are desired in finished goods inventory levels, backlogs of customer orders, or both and how quickly should they be made?
3/ For what manufacturing facilities will capacity plans be made?
4/ Over what period of time will the plan extend?
5/ What unit of measure of capacity will be used?

## FORECASTS FOR CAPACITY PLANNING

One of the basic characteristics of forecasts stated in Chapter 4 was, "forecasts are more accurate for groups of items." Since groups of items determine the load on manufacturing facilities, it makes sense to take advantage of this and use group forecasts for capacity planning. Such forecasts should, of course, be consistent with those used for profit planning and inventory planning. Statistical techniques, discussed in Chapter 4, can forecast a whole product family as if it were one item. To permit such forecasts to be reviewed by Marketing periodically, the families selected should be meaningful in the marketplace. If they are, the effects of competition and other external factors can be evaluated for each family, which is obviously not possible unless each family can be associated with a specific market.

Capacity plans, however, will be needed for groups of products made in specific manufacturing facilities, and these groups often do not match the product family groups which are best for forecasting. Conversions from forecast families to manufacturing families have to be made so that total demand on manufacturing facilities can be determined for use in planning capacity requirements. Here again the cardinal rule is, "get something practical rather than waiting until you can achieve a high degree of precision."

Differences between actual and forecast demand for product families should be checked regularly to see if the capacity plan is still valid. Such measures of error provide decision rules to use in deciding when to change capacity levels. Cumulative figures are better than period data for monitoring capacity. Figure 24 shows a good approach. The chart begins with the last significant revision to the family forecast and tracks cumulative actual against forecast. Limits of $\pm 10\%$ have been set as decision points. When cumulative actual demand exceeds the forecast limits for two consecutive months, capacity should be adjusted. Management judgment is needed in setting the control limits and in deciding how soon corrective action is needed.

*Don't wait too long!* The longer the decision is postponed, the more drastic will be the corrective action needed. You get hit two ways. Small weekly or monthly deviations add up fast to sizeable departures from plan. And the fewer the months left to reach planned levels of inventory or backorders, the greater the adjustment to capacity to achieve the change. This is illustrated in Table 7, which shows how much more the adjustment to capacity must be to compensate for a relatively small error if the decision to change is delayed.

## INVENTORY AND BACKLOG CHANGES

An important question on capacity planning concerns how quickly inventory levels should be adjusted or backlogs of orders brought into line. Inventories and backlogs are changed only by producing more or less than is sold. How much more or less depends on how quickly new target levels of inventories or backlogs are to be achieved. *Don't try to be a hero overnight!* Significant changes in inventory or backorder totals should not be made quickly. Keep the capacity adjustment small. Don't forget you'll have to alter capacity again once you get on plan, restoring cuts or reducing overbuilds so that you get output and demand back in balance and stay at planned inventory or backorder levels.

Attempting to change too quickly, which is really overreacting to previous failures to plan and control adequately, introduces very serious disturbances into any plant's operation and increases the difficulties of operating economically. The effect is much like flying an airplane. If it is allowed to drift off course for a long period of

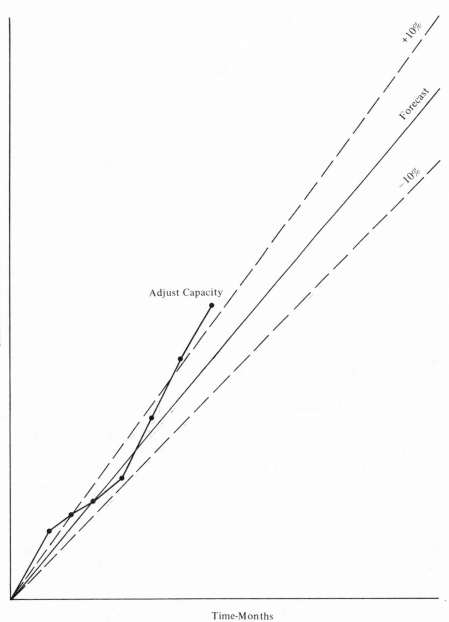

**Figure 24** / When to Adjust Capacity.

## Table 7 / Change Needed in Production after Delay

Forecast Shipments = $24 Million

Present T.O. = 2.4   Inventory Now = $10 Million

Target T.O. = 3.0     Inventory Goal = $ 8 Million

Target is to be met in 10 Months

10 mo. Shipments = $20 Mil     10 mo. Production = $18 Mil

Probable Forecast Error = 5% = $ 1.2 Mil

| Months To Goal | Change in Production | |
|---|---|---|
| | $ | % |
| 10 | 120 M | 7 |
| 8 | 150 M | 8½ |
| 6 | 200 M | 11 |
| 3 | 400 M | 22 |

time, a whole new flight plan is needed. Instead of returning quickly to the original course, it is better to plot a new course to the original destination, and to refigure the arrival time.

## SELECTING MANUFACTURING FACILITIES TO BE PLANNED

It may not be necessary to make specific production plans for all departments or work centers. Simple machining and assembly operations manned by housewives, college students or other temporary help can have capacity levels changed quickly and inexpensively to meet fluctuations in demand. Detailed capacity plans will be needed, however, where capital investment is heavy, where long periods of time are required to procure additional equipment, where skilled manpower with long training periods is needed or where stability of employment is desirable. Starting operations such as punch presses, forging hammers, mixing kettles, die casting machines, spinning and weaving machines and the like frequently establish the flow rate of materials through the balance of the operation, and are therefore the most critical in making accurate production plans.

Capacity plans for manufacturing facilities feeding each other must, of course, be consistent. Plans should first be made for finishing operations such as final assembly, packaging, painting, etc. Next, develop requirements for capacity in sub-assembly departments, parts fabrication centers, or processing operations feeding the finishing areas. In any case, the net change in inventories, work-in-process or backlogs of orders must be considered when developing the capacity needs. Between manufacturing operations, inventories of finished or semi-finished components can only be controlled by balancing the departments feeding and drawing from them. This planning sequence is easier to develop and follow where products have the same processing steps, for example, in chemical production, roller chain manufacture, and rug making.

*Get something started someplace!* Pick a bottleneck operation believed to limit total output, or one where loads appear to fluctuate widely, or one handling a self-contained product made complete—but get going if you haven't already. Work from the end of the process back toward the beginning so you can set output rates properly, and get work-in-process levels down where they ought to be. You need some measures of work content but high accuracy is not required. I'll come back to this later in this Chapter.

## THE CAPACITY PLANNING PERIOD

The planning horizon covered in a capacity plan must be at least long enough to establish the need for additional equipment or manpower. If it requires four months to procure, install, and get running a new piece of machinery, the need for it must be detected at least four months in advance. Likewise, hiring and training skilled people requires some minimum period of time which establishes the minimum horizon for the capacity plan.

It is obvious, however, that temporary peaks in demand should be met by overtime, sub-contracting or similar temporary expedients, and not by the purchase of equipment or hiring of permanent employees. The capacity plan, then, must be extended far enough into the future to determine if the indicated need for change will be permanent enough to justify capital investment or provide stable employment for permanent help. For capital equipment this usually means at least one additional year beyond the procurement time. For manpower the additional horizon depends upon a company's employment policy. Black & Decker, Dodge Manufacturing, McGill Bearing and many other companies would like to provide employment for every employee until he quits, retires or dies. This dictates that their capacity planning horizon extend as far into the future as meaningful numbers can be developed.

There is also a link between the planning horizon and the date to hit target inventory levels or backlogs of orders. The plan must extend at least this far into the future. There is obviously no need to have identical horizons in all capacity plans. Work centers easy to change could have very short plans, whereas those with more in-flexible capacity would have a longer horizon.

The most common approach is to make a "firm" quarterly plan showing the capacity levels required to meet the forecast total demand and to achieve the necessary inventory and backlog reductions or increases. Additional quarterly plans are prepared for one year out, updated at the end of each quarter. The firm plan is reviewed weekly and updated when significant variations from forecast or serious production problems make it necessary. The firm portion of the plan may extend farther into the future where the backlog of orders covers several months of demand or where manufacturing cycle times are long such as, for example, in building power plant generating equipment or large earth moving machinery.

## UNIT OF MEASURE

The unit of measure used for capacity planning should be the one most meaningful to those who have responsiblility for changing capacity. The best information is usually the hardest to get. Using standard machining-hours or man-hours requires considerable calculations to translate product forecasts into required capacity. It is surprising how frequently some simple measure, such as number of pieces, tons, gallons or similar easily obtainable unit, is satisfactory. Although frequently used for forecasting, cost dollars are not very good as a measure of capacity needed in most companies.

There are two problems to consider. First, the forecast must be converted to capacity units to establish the plan. Then,   actual output must be measured, and data in the same units compared with the plan. Practically every company has a labor-reporting system from which standard hours of work output can be obtained. Other reports may make it easier to tally up total pieces, tons, square yards or some similar measure of production output. The objective should be to *get some practical figures and get started.* It is not necessary that they be precise or even accurate, as long as both planned and actual figures are consistent. So long as you put requirements in and take actual out *in the same units,* it does not matter whether they are too large, too small or the correct size units.

## CAPACITY PLANNING TECHNIQUES

Techniques like that used to develop Table 6 in Chapter 5 are widely known and well-covered in the literature. They require only simple arithmetic plus some reasonably accurate—*not precise*—data on the work content of the product. A "bill of labor" like Table 8 for a typical product or for each product in the line fur-

### Table 8 / Bill of Labor

Model AB274        150 Units/Wk

| Work Center | | Standard Hours Required/Week |
|---|---|---|
| 110 | Assembly | 10 |
| 108 | Subassembly | 8 |
| 105 | Drills | 14 |
| 104 | Turrets | 28 |
| 103 | Mills | 9 |
| 101 | Stamping | 17 |

nishes more detailed information on the capacity required in the plant to produce it.

"Load" and "capacity" are not the same. Machine load programs have never been used successfully as capacity plans precisely because this difference has not been clear. Likening a manufacturing facility to a bathtub, the "load" is the *depth* of the water, and the "capacity" is the *rate* at which it's running in and out. Don't use a depth gauge when you need a flow meter.

Input/output reports, like Figure 25, have provided the first successful capacity planning and control tool. They can be set up for all kinds of operations. Planned input and output data are based on capacity required to meet the overall product output. If work-in-process is to be lowered, output must be jacked up to work off the excess. *Actual input and output are best determined by measuring actual flow of product.* This avoids getting embroiled with man-hours, efficiency, utilization and other confusing factors. Watching both input and output deviations pinpoints work-centers where the real capacity problems lie.

Very significant progress can be made in getting lead times under control and reacting quickly to demand changes, using fairly crude capacity plans. More sophisticated techniques are available using projections of planned orders and a computer to explode them through the routings and standards to develop detailed capacity requirements in all work centers but even these produce incomplete information. In addition, provision must be made for unplanned requirements such as scrap, rework, prototypes of new products, methods and standards changes, etc. The expense and effort of making the large number of calculations can pay off in better control. However, the bulk of the benefits can be obtained from simple capacity planning. I've seen work-in-process and lead times reduced by 40%, with better customer service and lower manufacturing costs, when one company developed capacity planning based on 19 "typical" products representing their full line of over 650 items.

## CAPACITY PLANNING FOR PURCHASED MATERIALS

Very successful use of the capacity plan is being made in many companies for purchased materials. They purchase capacity from a vendor to insure more dependable deliveries of specific components. This works particularly well where the vendor supplies a number of items produced on the same manufacturing equipment. In concept, it treats the vendor exactly like a department of the customer's operation. For example, ball bearings, electric motors, screw machine and sheet metal parts, castings, forgings, bottles, wire, and many other items are purchased with lead times of four weeks or less, coupled with a commitment to take the output from a planned amount of capacity over four or five months. When

| Week Of | 3/06 | 3/13 | 3/20 | 3/27 |
|---|---|---|---|---|
| Planned Input | 310 | 310 | 310 | 310 |
| Actual Input | 205 | 260 | 270 | 250 |
| Cumul. Deviation | -105 | -155 | -195 | -255 |
| Planned Output | 310 | 310 | 310 | 310 |
| Actual Output | 240 | 210 | 260 | 280 |
| Cumul. Deviation | -70 | -170 | -220 | -250 |
| Actual Backlog | 520 | 570 | 580 | 550 |

**Figure 25 /** Input / Output Control Chart (Standard Hours).

capacity is purchased, the vendor knows what manload levels and equipment will be needed, what raw materials will be required, and he can plan better to meet all his customers' needs. On the other hand, he does not need to know specifically which individual items the customer wants until just before he's ready to start production. This approach is far more economical and much more effective than "blanket orders" for a specific item, or having the vendor carry the items in his inventory.

*Purchasing capacity is really the only practical approach to getting purchased materials on time.* Periodically our economy goes through "lead time" cycles. An increase in demand, sometimes significant but often very small, builds up customer order backlogs because manufacturers don't recognize or can't react to the need to increase capacity quickly. The standard, accepted, inevitable reaction is illustrated in Figure 26. When schedules "fill in," obviously lead times must increase. Or must they?

The fallacy of this is in *attempting to bring planned average lead times and actual average lead times back into agreement by increasing the planned averages.* It can't be done! All this really does is cause customers' ordering systems to trigger the release of more orders supposed to cover requirements further in the future. The result is larger backlogs and less reliable delivery dates (remember, "forecasts are less accurate far out"). Capacity is wasted (working on items the customer ordered long ago but really doesn't need now) at the very time it can be least afforded.

## CAPACITY CONTROL IS VITAL

Without effective planning and control of capacity, your company will always be in some kind of trouble. The alternatives are all unpleasant: either too little total output, which means some customers (hopefully not the best ones) will be unhappy about missed deliveries, or too much, in which case you can build excess inventory or waste manpower and equipment capacity.

All our planning activities hinge on dependable lead times. We've discussed three different kinds:

1/Planned average lead time—what we've told the ordering system it should take to get materials, *on the average.*

2/Actual average lead time—what it actually takes to get materials through all or any part of our plant or our vendors', again on the average.

# BEEDEE FOUNDRIES, INC.

Chicago, Illinois

Executive
Offices

June 1, 1972

Mr. J. E. Smith
Reliable Vent Company
100 East End Road
Minneapolis, Minnesota

Mr. Smith—

Many purchasing people have been asking for an update on lead time for casting procure-ment. This is, of course, a difficult question that varies week to week.

The important point for you to consider, however, is that the business tempo is increasing and schedules are filling in. Your help is requested in placing firm orders for the coming three to six (3-6) months.

Since the plants start cores, and in some cases molding, as much as 60-90 days prior to scheduled shipment, you are asked to be realistic in your schedules and consider the first ninety (90) days of any schedule as a firm commitment.

Now for the information that you requested. Twelve to fourteen (12-14) weeks is the minimum time you should be working with and when you are scheduling with computer, use 16 weeks.

Thanks for your help.

Sincerely,

William Stone
Vice President
Sales and Marketing

WS:pm

**Figure 26** / Beedee Foundries Letter.

3/Actual specific lead time—what we can really do if we give proper priority to an individual order.

Capacity planning and control are needed to make the average lead times—planned and actual—keep in step. *You must change the actual average by controlling capacity to get them into agreement.* You cannot do it by altering the planned figure. Only when these two averages agree can you expect to have things like customer service, inventory levels, and manufacturing costs come out where you want them.

The third lead time—what any single item might take—is controlled by the priority system. How to make sure you know which item has what priority is the subject of the next chapter.

# 7

# Making the Right Things at the Right Time

The fifth and last basic question to be answered is, "Are we working on the right items today?" Although it's the last activity in the sequence, controlling priorities is not the least important—and it's by no means the easiest. Too often it catches the eye of management as an area to be attacked first. It's tempting to think that a good system of identifying material and its location in the plant and collecting information on the status of all work would really "get out the things we need now." Unfortunately, it rarely helps—until priority planning and capacity planning and control are effective. There are no short cuts!

*First, you need a good ordering system to tell you what materials you really need and when you need them—and to keep this information current.* Priority control begins with priority planning all the way back to the master production schedule. This master schedule provides input for both priority and capacity planning. When making capacity plans, start with a master schedule of *what you want to make* so the plan will develop *the total capacity you would need* to make it. When planning priorities, however, you

121

have to be realistic and start with a master production schedule containing only *what you are capable of making with your available capacity.* Lie to the system about this (everybody does, at least once) and it will lie back to you on the priorities of material requirements.

*Next, you must have adequate capacity or knowledge of how much you're short if total output is inadequate.* Priority control depends on managing lead times properly, both the average lead time of many items and the specific lead time of any one. Within available capacity, input control then takes over to get things started right.

Input control techniques are needed to level out the flow of work to the plant, to release it in a balanced mix which the plant can handle without bottlenecks or downtime. These techniques include selecting the proper jobs to achieve this balance, scheduling them to set milestones so that their progress can be checked as they move through the factory, and developing work center loads to indicate overloads and underloads which will require rescheduling jobs the facilities cannot handle.

One of the cardinal principles of production control is that *work input should be less than or equal to actual output, never more.* Its objective is to prevent the buildup of work-in-process, lengthening average lead times and making individual lead times less predictable. It is probably the most difficult principle to implement. The common reaction is, "orders must be released when the inventory planning system says so." Also "the factory must be given the orders if we want work out on time." Both of these have already been shown to be wrong. The inventory planning system is designed to tell us when we should release specific orders, based on the ground rules we gave the system: master production schedule, balances on hand and on order, and standard lead times. Releasing the order when the inventory system says so is worse than useless if the plant is already loaded and can't handle it. That's like pouring faster to get the contents of an 8-ounce bottle into a 6-ounce glass. A sound and disciplined input control system is vital to control lead times and get the right priorities established.

The principle states that input should never exceed *actual* output. Notice it is actual, not planned output that should limit the input. There is frequently a problem in deciding on the best way of determining what the output capacity of a plant really is. Manufacturing engineers can figure in detail what the capacity should be based upon nominal manload and existing equipment, assuming certain shift operations. These calculations are difficult to make: a man can run one, two or three machines; during a partially manned shift, men may be switched among different machines; changes in product mix result in different total amounts of output, and many factors such as absenteeism, injury, union meetings, etc. reduce the plant output below the calculated level. For input control purposes it is not necessary to get involved in all these details. *Simply replenish each week the amount of work a work center or department turned out last week.* This is the best measure of actual capacity. This simple approach works very well; the real difficulty is in getting people to believe it can.

The inventory planning system should provide a list of orders from which to select a number of jobs to be released to starting and assembly operations. An input/output control plan similar to Figure 25 in Chapter 6 is a very useful tool to monitor actual output compared to planned levels, and to insure that the right amount of work is being released. Select a unit of measure which makes it easy to total up output and to match it with new orders. If a starting operation also handles secondary operations on material started in another work center, this additional work load should be shown separately in the planned and actual output, and measured in the actual input. New-work input figures should match new work completed. The purpose of input control is to regulate the release of orders so that work-in-process levels stay down where they should, and the control of priorities stays with those doing the ordering, not with those sorting out the backlogs in the plant.

Although the flow of work to secondary and intermediate operations cannot be controlled as closely as starting and assembly operations, an input/output plan should be developed for these also. It provides an excellent tool for early detection of failure to produce up to planned levels, and identifies clearly for each work center whether or not input and output are adequate.

## BALANCING THE MIX

It is not enough to feed the correct total of work to a work center; there must also be a good balance among individual operations. Work centers with a variety of sizes or types of equipment and with little flexiblity to shift people to different operations must have some work released to each group. In such operations it may be better to set up individual input/output plans for each of the subgroups. However, improving the flexibility of labor and equipment can pay real dividends not only because it reduces detailed planning but, more important, it is faster to respond to changing priorities.

Potential bottlenecks or downtime in intermediate or secondary operations can be headed off by releasing the proper balance of work in starting operations. Coining press operations were a problem in one forge shop. Some parts required six times as much flattening and sizing operations as others but they all looked alike to the forging hammers. Consideration of these coining requirements when the orders were released to the forging operations made it possible to feed in a balanced mix which gave a smooth flow of work through the coining presses. Identifying critical factors like these, anticipating bottlenecks and planning better work flow can minimize the difficulties of controlling priorities after work is started.

An input/output control chart works just as well for purchased materials as for manufactured. Fisher Controls Company in Marshalltown, Iowa, used the chart shown in Figure 27 for purchased castings. The foundries who deliver their full output quotas receive new orders equal to their commitment; those who fall behind schedule find that the input of new orders also falls off. Fisher thus follows the basic principle of keeping input equal to output. Coupled with short lead times (four weeks), this gives a real incentive to foundries to stay on schedule. It works far better than the old approach of more lead time and higher pressure on suppliers to get materials that have not been delivered on time.

## SHORT-CYCLE SCHEDULING

Control of priorities is easiest if only small amounts of work are released at a time to the plant. If a month's work is released, the plant will have a wide choice of what to start first. The smaller the amount of work released, the narrower the choice and the easier it will be to control priorities. Of course, there is a trade-off; more frequent scheduling means more work for schedulers, so the number of people and their system's ability to handle the work load

PURCHASED CASTING AGREEMENT
FISHER CONTROLS CO. AND
XYZ FOUNDRY

DATE: August 1, 1972

PRODUCTION LINE: All

| Month | Past Due | | Released Orders | | | | Capacity Committed | | Forecast | | |
| --- | --- | --- | --- | --- | --- | --- | --- | --- | --- | --- | --- |
| | Prior | Prior | July | August | Sept. | Oct. | Nov. | Dec. | Jan. | Feb. | March |
| Requirements | X | X | 760 | 860 | 760 | 950 | 760 | 500 | 760 | 760 | 760 |
| Cap. Allocated | X | X | 760 | 860 | 760 | 950 | | | | | |
| Cum. Deviation | X | X | X | X | X | X | | | | | |
| Ordered | X | X | 729 | 801 | 768 | 945 | 760 | | | | |
| Received | X | X | 37 | 10 | | | | | | | |
| Cum. Deviation | X | 376 | 1068 | 1859 | | | | | | | |

Next Month's Release → (Nov. 760)

Note: December capacity commitments reduced because of poor deliveries.

Signed:

Buyer – Fisher Controls Co.

Vendor – XYZ Foundry

**Figure 27** / Purchased Casting Agreement.

must be considered. Most well-run plants are run with weekly schedules, but some are using daily schedules to improve priority control. Examples of these will be given later in this chapter.

## AVAILABLE WORK

Many managers are reluctant to operate "lean" plants, with low levels of work-in-process. They think they need "cushions" to cover planning mistakes and other upsets. Record errors, tools not yet available, missing raw material, parts held up for engineering or quality questions are familiar happenings. Scheduling systems, of course, should include a review of all ingredients necessary to get work started and keep it going: tooling, material, incentive labor standards, special setup instructions, blueprints, job tickets, etc. Work held up for engineering or quality questions should be identified promptly and those responsible for getting answers alerted.

Higher levels of work-in-process are not a valid alternative to fast response and tight control. The real choice is between running a priority system with planned lead times which are reliable and solving problems, which in turn destroy the validity of priority planning. There can be little doubt that an effective system will require people to detect problems and solve them promptly.

## SCHEDULING

There are three purposes in scheduling:

1/ To establish dates on orders as milestones to measure progress through plant operations and determine whether they are on time, late or early.

2/ To aid in setting relative priorities among orders competing for capacity at a work center.

3/ To help determine probable short range loads in work centers.

There is a widely held belief that scheduling requires accurate, complete, detailed routings, time standards on all operations, and a computer program to make the necessary calculations. This is patently ridiculous. The task is to develop an estimate of lead time for each operation to be scheduled. Lead time in manufacturing is made up of paperwork time, preparation (including setup) time, processing time, move time and queue (or delay) time. To develop a schedule of orders moving through the plant, some estimates must be made of each of these elements. It is possible to calculate preparation and processing times and make close estimates of paperwork and move times. However, in all but a very few plants, the sum of all of these elements of lead time will be no more than ten percent of the total. Queue times, delay times caused by backlogs of orders, account for the bulk of most lead times, and there is no practical way to make detailed estimates of this element. When this is recognized, it becomes obvious that precision is not the name of the scheduling game.

With routings, time standards, some scheduling rules and an open order file, a computer can schedule a large number of orders and update the schedules regularly, daily or weekly. There are two approaches. Forward scheduling starts with today's date and develops the specific date when each order will move into or out of the work centers involved. Backward scheduling begins with the date an assembly or component part is due in stock and works back through the operations required to the time it should be started. Forward scheduling is used with detailed shop loading to find out if the earliest feasible completion dates will meet the customers' requirements. Backward scheduling is used with assembled products in which the component lead times vary widely and it is unecessary to start all of them simultaneously. It develops the latest start dates for the components.

It is very tempting when setting standards for the elements of lead time to "play safe," putting cushions in each element to "be more certain of meeting the schedule." The effect of this padding, however, is to force the planning system, because of the long lead times, to release more work to the plant. This builds up backlogs, increases the work of developing schedules, and makes it more difficult to keep priorities straight. Lead time standards for scheduling purposes should be realistic but short.

Using the A-B-C approach identifying the important few jobs contributing the bulk of the total load or critical in meeting specific delivery requirements, manual scheduling techniques can be very helpful. Another shortcut is to schedule only one or two critical operations or work centers which control the flow of work in the plant. It is possible to achieve 80% of the results with 20% of the effort. It's a serious mistake to delay starting some scheduling activities for want of detailed, precise systems.

## MACHINE LOADING

As I said back in Chapter 1, machine loading is the second oldest inventory control technique, but has rarely been a useful one. We've known how to make work-center and machine loads since Fred Taylor showed us how to set work standards and Henry Gantt taught us how to draw bar charts. Because of the confusion between load and capacity which I discussed briefly in Chapter 6, machine loads are usually applied as capacity plans, and usually fail.

Most machine loads use only firm orders already released. This places serious limitations on the amount of information in the load plan for future requirements. The horizon is far too short for effective capacity planning. Another major problem is that the orders released are generated by poorly designed, incorrectly applied ordering systems, and the required dates are anything but accurate. In addition, scheduling systems are often inadequate, being based on inaccurate dates and being so unwieldly that they are impossible to update as changes occur in the flow of work. For these reasons machine loads have been almost completely ineffective. It's usually much easier and more practical to determine plant loads simply by walking out on the floor and looking at them than it is by studying the machine load data.

*Machine loading is a priority control technique, not a capacity control device.* Its function is to illustrate *in the short range* the effects on individual work centers of the orders scheduled to pass through them. In this short range period it can develop reasonable measures of overloads and underloads which will occur if the work proceeds on schedule. This information can then be used to make decisions on short range corrective actions to alter capacity so that the schedule can be maintained. These corrective actions may include transferring people from lightly- to heavily- loaded centers, subcontracting, overtime, alternate operations and other temporary or interim expedients. If such corrective action cannot be taken, it is obvious that the schedule cannot be met and priorities will have to be changed so that available capacity is utilized for the most important orders.

There are two distinct approaches to machine loading. "Loading to infinite capacity" assigns the work content of all orders to individual work-centers in the time periods specified by the schedule. It pays no attention to the available capacity in that work center. However, when average load figures are developed with this approach, using *planned orders* well into the future in addition to firm released orders, load data can be used for capacity planning purposes. In fact, we have recommended that machine loading applied in this manner be called "capacity requirements planning," and this is now accepted.

The second form is called "loading to finite capacity." It assigns work to work-centers in priority sequence so that the highest priority orders get first claim to available capacity. When the limit of available capacity is reached, orders will have to be rescheduled earlier if capacity is available, or dropped back into later periods until capacity to handle them is found. Very few companies have applied this technique successfully as yet, but it has some promise as a finetuning, priority control device. Obviously, it cannot be effective until a sound ordering system is available to provide valid due dates, until capacity is adequate to meet the average load, until reasonably accurate standards are available for measuring the work content of individual orders, and until a computer program is operational to handle the masses of calculations involved.

## PRIORITY CONTROL

Input control attempts to get work started properly, in the right priority and with the greatest chance of getting finished on time. Many things occur, however, to interfere with planned progress of work. Some techniques are needed to detect and initiate corrective actions for such upsets as revised customer delivery dates, forecast changes, vendor falldowns, equipment breakdowns, employee absenteeism, tooling problems, engineering difficulties, quality rejections and the like.

The most effective priority tool is the due date on the order. Priorities will be meaningful, however, only when the due dates are valid—maintained up-to-date, with changes. The order point/order quantity system needs supplementary techniques like Critical Ratio to do this, but material requirements planning will keep due dates current as the master schedule is changed or as replenishment orders progress or are delayed. While it is possible to use "Time Remaining" (days left until due date) as the priority ranking device, it is evident that some measure of "Work Remaining" (days required to finish the work) should be considered also. The ratio of these two, called Critical Ratio, is a common priority control technique, and there are several others. A good priority technique should:

1/*Be valid,* based on realistic due dates, kept up to date.

2/*Be dynamic,* changing priorities to reflect both customer-and factory-oriented changes.

3/*Give relative priorities* to the maximum number of orders; have as few "ties" as possible.

4/*Be simple,* since it will be used by almost everyone and should be easily understood.

Many managers associate priority control with data collection equipment. While the equipment itself is useful, it contributes only two minor benefits to priority control:

1/Fewer errors—by eliminating some manual data entry, particularly handwriting and keypunching errors and by editing data.

2/Faster collection—by converting input data directly into machine readable form and speeding transmission from source to system.

Some data collection is highly sophisticated, capable of guiding the individual entering data through the proper steps to insure that complete information is collected, and equipped with edit programs to test data entered for validity on a large number of characteristics. On the other hand, one of the most effective data collection techniques we know is a man on a bicycle who collects job cards daily and delivers them to Data Processing for entry in the system.

Data collection and priority techniques do not constitute effective priority control. This requires that five activities be handled properly:

1/*Ranking orders in desired priority sequence* indicating the next order to be run, the one after that, etc. The most commonly used tool for this is the "dispatch list" updated daily or at least weekly. Figure 28 shows the daily dispatch list used by Twin

Daily Work Center Job Schedule

| Date 07-13-70 | | | Machine Center BH | | | | Week 315 | |
|---|---|---|---|---|---|---|---|---|
| Part | | Priority | | | | Next | Remaining | |
| No. | Name | PO | PI | Qty | Hours | Loc. | Work | Time |
| 6563A | Piston | .578 | | 332 | 17.9 | 0103YE | 24.1 | 14.0 |
| 3777 | Flt Plt | .698 | | 28 | 4.2 | 0103HE | 20.0 | 14.0 |
| | • | | | | | • | | |
| | • | | | | | • | | |
| | • | | | | | • | | |
| 3471AH | Sleeve | | .061– | 11 | 3.8 | 0103Y8 | 16.2 | 1.0– |
| 3552M | Cone Slv | | .074– | 13 | 3.8 | 0103Y8 | 13.4 | 1.0– |
| | • | | | | | • | | |
| | • | | | | | • | | |
| | • | | | | | • | | |
| Total Hours This Center | | | | | 268.7 | | Capacity | 257.2 |
| Parts in Previous Work Center: | | | | | | Previous | | |
| 3471M | Sleeve | .781 | | 300 | 12.5 | 01068CA | 17.9 | 14.0 |
| 6751 | Press Pl | | 1.076 | 275 | 13.6 | 01038H | 17.6 | 19.0 |
| | • | | | | | • | | |
| | • | | | | | • | | |
| Total Hours for This Machine Center in Previous Center | | | | | 236.0 | | | |

**Figure 28** / Daily Dispatch List.

Disc, Racine, Wisconsin. It shows orders ranked by priority in two groups: material for customers (PO) and for stock (PI). The ranking is based on the (Critical) Ratio Time Remaining/Work Remaining, and these factors are shown along with the ratio. The listing also shows the next work center for each order, as well as new orders expected soon and the work center they are coming from. The list is issued daily at the start of the first shift for each work-center, based on order status at the close of the second shift the preceding day. Space is left at the top for the dispatchers and foremen to enter additional orders or to rearrange the priority sequence if they are aware of factors not known to the priority system. This is a well-disciplined, well-managed priority control tool.

2/ *Provide information on subsequent orders* so that needed preparatory work can be planned and carried out. Pre-cutting raw materials, cleaning or adjusting equipment, rearranging work areas, and many other activities can be handled better if planned with a knowledge of expected as well as existing orders.

The common approach is to include several day's or a week's work on the daily dispatch list to give the needed visibility of future orders.

3/ *Develop information for lot control* including job location, consistent counts on succeeding operations, and cost data to compare with standards. Lot control can be based either on discrete order quantities or distinct time periods, the latter being used when operations are continuous over several days or weeks.

4/ *Furnish timely and accurate feedback* on plant activities not proceeding according to plan, and on order status and location. This is intended to keep the priority control system informed of how plant activities are progressing.

5/ *Reschedule activities* as necessary to keep the plant abreast of changes in priorities. This is intended to keep the plant up-to-date on the latest priority information.

These activities must and will be performed, either promptly and effectively using a well-organized, disciplined priority control system or on a "too-little, too-late" basis using informal, crisis-controlled expediting. Costly, complex data collection and priority techniques are not only unnecessary to make real progress, they will be just so much wasted effort unless applied in connection with good priority planning and capacity planning and control. Painting the superstructure and polishing the brightwork is useless if the sails are torn and the hull leaks.

# 8

# Developing
# Better Systems

## Can You Afford the Best?

### WHY NOT INFORMAL SYSTEMS?

All companies have a formal manufacturing control system of some sort, although finding out what it is can be a real challenge. There are usually at least four versions: first, the way the system is described in its documentation (if any exists) but, usually ancient, such a description bears little resemblance to the real system; second, management's idea of the system, which also bears little resemblance to the real thing; third, the system as understood by all those using it and thinking they understand how it really works, but each with a different interpretation since each person sees only a part; and fourth, the actual system, which can be understood only after much investigation.

In only a very few companies is the formal system capable of really controlling the operation. Most formal systems can be characterized as "providing the paperwork to authorize what people decide to do based on what the informal systems tell them."

133

Spotlighting the failures of formal systems are the many useless reports generated periodically and ignored continually.

The system, defined in Chapter 1, is the structure of records, transactions, reports, and decision rules providing the means for implementing planning and control activities. Its purpose is to assist those controlling operations through prompt detection of significant deviations from plans and timely reporting of the need for corrective action. Systems for manufacturing control are information-based and materials-oriented.

Since most formal systems are impotent, they must be supplemented with informal systems. Larry Wightman of the Emerson Electric Company, said, "Innovation is born of frustration and adrenalin—not love," and although he didn't intend it, he might well have been describing the origin of informal systems. Humans are distinguished from animals by their ingenuity in developing and using tools to enlarge their physical and mental abilities, and in adapting to their environment and the conditions it imposes on them. Nowhere is innovation, ingenuity and adaptability more evident than in the informal systems developed to assist people who do not have the proper formal tools to do the jobs they have to do.

The most conspicuous example of an informal system is the "hot list" developed by an assembly foreman, which lists the components he really needs to put the product together. Although really too little, too late, these hot lists are usually the only true priorities. A staging area, where materials are set aside to be sure all components are available before orders are released is another prime example. An overworked informal system is on-the-spot counting to be sure the true quantity and identity of critical materials are known. Other examples are the side-records found in the foreman's "black book," and hand-written or memorized bills of material describing how the product *really* goes together. In all cases, these sub-systems are accurate measures of specific failures by the formal system to do its job.

Without these informal systems, most companies' operations would be out of control. In times of unusual stress caused by rapidly increasing or decreasing product sales rates or major changes in the product line, informal systems become overburdened and are incapable of control. The resulting crises are all too familiar in companies dependent on informal systems. Recovery depends on how quickly and well people react.

All systems are simply tools of the people who use them. We need people and systems working together, but we must develop much less dependence on people and more on systems if control programs are to be effective. The objective is not to replace people but to make better use of their capabilities. Systems must relieve people of routine tasks, presenting information for decision making and exercise of judgment. Informal sub-systems must be eliminated simply because they do not do this well. Systems should manipulate; men should manage.

If there is a single word describing the results of manufacturing systems development efforts it is "frustration." The feeling is prominent among all levels of management. Top management is frustrated by the combination of high costs and unfilled promises of computer systems. The feeling is aggravated by problems in communicating with specialists and evaluating their contributions. They read the "success stories" (mostly fictitious) of other companies whose systems are supposedly working well and see little chance of this happening soon in their own operations. They face the apparent need for long and costly programs with no real assurance of success at the end. They have little faith in the prophets of the great god "System."

Middle management, particularly in manufacturing, is frustrated by the conflict between what they think they need in systems and what the data processing specialists say they should have. The fact that both are wrong merely accentuates the conflict. The pressures of fire-fighting to meet daily problems militate against taking time to develop needed fire prevention programs. The conflict of priorities between what line managers consider "scorekeeping" and the systems they want to "help the players" adds to their frustration. Long delays in getting something running are all too typical even after agreement has been reached. Line managers are their own worst enemies, however. People with experience in trying to interest busy operating managers in new systems can undoubtedly add to the "Eleven Ways to Kill an Idea" listed in Figure 29.

Many refuse to be led to water, let alone to drink at this well. Hardly anyone is more frustrated than systems and data processing people. Typically intelligent, ambitious and aggressive, they see the great potential of proper applications of computers and sound systems. Unlike the typical user, they spend a fair proportion of

Are you pestered by idea men? Have you longed for an effective way to get rid of them? Do you resent attempts to drag you out of the status quo? Rest easy. Here is a foolproof set of rules for dealing with do-gooders and their rest-disturbing ideas.

LET'S ORGANIZE what we know about *ideocide*. Daily, we have to kill ideas that may develop too far and cause us to think and work along unfamiliar paths. Here are the 11 most successful methods used to quash concepts and throttle thought. Do not try to think of other ways. It doesn't pay. They've probably been tried before, won't help, or are a waste of your precious time.

**1** **Don't be ridiculous.** After all, if a new idea sounds absurd, why continue? If someone else thought of it first, the odds are that the idea is ridiculous anyway. (This concept is also known as Quasher's Law, named after Dr. Emil Quasher, who, as new products director for a large conglomerate, has so far turned down xerography, the hula hoop, microcircuits, and radar.)

**2** **We tried that before.** Beware of these defenses—"It was not used in the best way previously." "A new means of utilizing the idea has been thought of." Reply that it is just wasteful to repeat a past failure. Bellow loudly that no upstart could possibly think of a new way to use an old, probably ridiculous idea.

**3** **We've never done it before.** Simple. If it has not been done before, there must be a good reason for not doing it now. Think of all the extra time, effort, and money that will have to be mobilized to do something new.

**4** **It costs too much.** Watch out, someone will say, "The final results will justify the costs involved." Don't be fooled; you know, deep down, that no one can know whether the results will justify anything. (Also known as Finagle's Principle, after Mrs. Rhoda Finagle, famous anthropologist, who, during her stay with a now-defunct group of Blund Islanders, convinced the natives that they could be more prosperous if they went back to yam farming and stopped making transistors for Japan and the U.S.)

**5** **It's too radical a change.** Almost self-explanatory. The wording may be unfamiliar, but it's an old concept. "Don't rock the boat." "Take it easy."

**6** **We don't have the time.** Face it: if you wanted to work more hours, you would have thought of the idea yourself—or got a second job. It takes enough time to merely deal with the old things.

**7** **We're too small for it.** Maybe, when we get to be bigger, we can handle it, but not now. Do not let yourself be convinced by the argument that this idea may help you to get bigger. You can't be sure of that.

**8** **Top management would never go for it.** You know how top management feels about things. You wouldn't be where you are if you didn't. Maybe, when you are part of top management, you will be more receptive to the idea.

**9** **We'll be the laughing stock.** Be sure to say this quickly—without pausing. Continue by changing the subject. It is imperative that you do not let anyone continue what you have said with "All the way to the bank." This is not a joke, you know. Be careful of talk of innovation, profits, invention, or creativity.

**10** **Let's form a committee.** Most-used way to kill an idea. After all else has failed, if an idea continues to survive, use this. You know the classic story: an elephant is a jackass put together by a committee. Set up a committee to study the first committee. Set no deadlines. Outline no procedures. Let the committee go its own way, so that a thorough, proper study can be made.

**11** **It's not my job.** This may be the best way. If it is not your job, why should you do it, or even listen to it? Do not be fooled into doing someone else's work. □

**Figure 29** / Eleven Ways to Kill an Idea. (Reprinted by permission from Industry Week, June 7, 1971.)

their time in educational programs expanding their knowledge and understanding of available techniques. However, they are too easily oversold on the benefits of mathematical and computer solutions, and too little aware of what's needed to apply these effectively to the real problems of operating a business. Some of their frustration stems from what they regard as too much emphasis on "people problems," and some from operating managers' failure to get the necessary disciplines to make systems work after they are designed. They are impatient with the idea of evolving systems from simple to complex, preferring to move directly to the "grand system."

When we don't understand a situation, we tend to oversimplify or overcomplicate it. Systems development efforts are no exception. They have been characterized by Instant Cures and the Total Integrated System. Many managers are beguiled by instant cures, possibly as a result of watching too many TV commercials about pills for curing headaches or chemicals for opening clogged drains. These managers try to cure sick profits and move sluggish inventories by speedy application of packaged programs and other "bandaid" approaches. Unfortunately, these too frequently attack only the symptoms, not the real diseases. For example, the real problem in most companies is not to get an inventory record or bill of materials loaded on a computer but to get an *accurate* inventory record or *properly structured* bill of materials. The "instant cure" type managers hope to avoid or don't see the need for an organized continuing effort for systems development. If one instant cure doesn't work they usually have another one ready and waiting.

The complicators think that the Total Information System is the only right way. They look forward to tying all company records and activities together—linking sales, engineering, manufacturing and cost accounting data—in a massive system. To date, this has provided excellent excuses for postponing doing anything while the system which will do everything is being designed. The Total System has not yet happened and there is no evidence that it ever will.

## HOW SICK ARE YOU?

The procedure for effecting a cure for sick systems is now fairly well recognized. The first step is to get specific measures of present system deficiencies. This accomplishes two things: first, it indicates where you should begin work (obviously on those elements which are weakest); second, it provides a measurement base to determine whether or not you are making progress in effecting the cure. Today's problems always seem the worst we've had unless we have some specific measures to prove that yesterday's were even worse. Here are some symptoms you should consider:

1/ The percentage of items backordered, the percentage of items out of stock, the number of customer complaints or other measures of customer service you have been accustomed to using or think are pertinent.

2/ The percentage of manufactured and purchased orders past due, how many are really needed now, plus other measures of the reliability of the ordering system.

3/ The value or number of undelivered requisitions in the stockroom for which there is no material available.

4/ Average actual vs. average planned lead times, plus minimum and maximum lead times on families of items.

5/ Production output, both total and for critical work centers.

6/ Number of items on assembly hot lists or expedite reports.

7/ Number of errors in stock status reports, bills of material, shop routings, open order files, assembly schedules and other important reports.

8/ Total inventory versus budget, by major product lines or major inventory classes.

9/ Significant expense items such as downtime, overtime, sub-contract premium costs, extra freight and other items associated with poor planning and crises.

Some new goals should be set for improvement in these indices. A simple technique is to take the best figure in any time period and make that the average goal for the future. Or you can aim for a percentage reduction which seems reasonable. This is an area where the experience of other companies is helpful in setting realistic goals. Providing such data is one of the real contributions consultants can make. They are happy to furnish information on gains that their clients and other companies have made. These should be checked out thoroughly, of course, and not taken at face value.

## WHAT IS THE PAYBACK FROM BETTER SYSTEMS?

*The only reason for better systems is improved return on investment.* Significant payback can come in each of the major areas of customer service, inventory investment, and economical operations. Figure 30 gives a checklist of specific factors which

Inventory Investment (15-30% reduction)

    Stocked Components
    Safety Stock
    Lot Size Remnants
    Obsolete
    Unique vs. Common
    Finished Product
    Service Parts

Operating Expense (2-10% reduction)
    Set-up and Change-over
    Unplanned Overtime
    Sub-contract
    Downtime
    Indirect Labor
    Productivity
    Warehouse, Taxes, Insurance

Customer Service (consistently at desired levels)
    Sales
    Freight
    Substitutes
    Back-Orders
    Clerical, Mail, Telephone

Prompt Planned Response

Coordinated Management

**Figure 30** / Benefits Checklist.

could be evaluated for your company. It also shows some typical benefits which companies have achieved by improving control of operations.

The job of estimating specific potential payback should be assigned to those managers who will be responsible for making the savings. Never turn this over to staff or systems people or outside consultants. The real problem here is in striking the right balance between ultra-conservative figures which represent no challenge, and exuberant estimates which probably will never be achieved. It is better to err on the conservative side, however, since this eliminates potential excuses for not making the goals by managers who are looking for them, and prevents the whole program of systems development from getting a "blue sky" image. You can afford to be conservative. Whatever estimates you come up with, you will quickly recognize that *there is nothing better you can do to improve return on investment.* We have yet to see a company which did not give top priority to this type of program after making a detailed estimate of potential benefits. Also, we have yet to see a company where the savings did not pay off the costs of systems development in less than one year.

Some factors to include in these cost estimates are listed in Figure 31. Almost invariably, reduced inventory investment and

Team Members
    Full-time
    Part-time

Clerical
    Full-time
    Temporary

Overtime
    Salaried
    Hourly

Travel

Hardware
    New
    Temporary

Software Packages

Consulting Services

Education
    Outside
    In-house

**Figure 31** / Systems Budget Items.

improved labor productivity together provide all the return needed to justify the systems development investment. However, don't overlook other potential savings. If you do not make specific plans, you won't get them.

Don't put up with long projects. Look at the potential monthly savings and it will be obvious that you can't tolerate long drawn-out projects. You can afford to spend a lot to avoid deferred profits. One company revised a twenty-one month project into a six-month plan because the general manager just would not tolerate waiting the additional fifteen months for the benefits to start. (The extra costs for collapsing the timetable were paid off in only two months.) Another company converted a forty-month project into a twelve-month plan for the same reason. The chances of successfully completing any project taking longer than one year are about as good as snow in Georgia in July. Something inevitably happens, like major management changes, acquisitions, new plants or some other crisis, which causes efforts to be diverted into other activities. If the benefits are really good, why not get them as soon as you can?

## ORGANIZING THE EFFORT

Responsibility for directing the program should be assigned to a manager, preferably the one responsible for achieving the bulk of the savings. Because of the nature of the program, a team will be necessary to do the work. This does not mean a committee to "vote" on detailed decisions. It takes a project team of representatives of production control, manufacturing, product engineering, data processing, marketing, cost accounting and other departments, some full-time and some part-time. Each department representative should make the detailed decisions on program requirements for his department. The ideal team is made up of those who will be directing operations after the system is installed.

Since there is no other program which will have greater influence on the future profits and success of your business, the project team should be the best you can put together. Don't select men who "can be spared," old employees near retirement, or young people who have just come aboard. There is no harm in selecting a manager new to your operation if he has experience and is going to play a significant role. To get the best job done quickly, make it a full-time assignment for key people, including the production control, data processing, and manufacturing representatives. Assign top priority to the program and stick to it. Don't let firefighting activities interfere. If the benefits are really that good, what else can be more important?

Outside consultants have a role to play in such programs, but it is not to do the job for you. What a consultant can do to make a real contribution is to:

1/ Provide objective analysis because he is an outsider.

2/ Help you avoid the mistakes of using the wrong tools or failing to recognize their weaknesses because he is familiar with available techniques and their successful application.

3/ Help you avoid expensive pitfalls, and show you successful applications by being familiar with what other companies are doing.

4/ Help you to evaluate costs and potential paybacks, and to set realistic schedules.

5/ Use his periodic visits as deadlines for the completion of specific assignments, and spurs to keep the project moving.

6/ Provide assistance in conducting educational sessions for all those involved in the design, installation, and use of the new system.

In summary, a consultant should be looked to for advice and counsel, guidance and direction, information and education. The basic decisions should be made by your people. A sign of real trouble is having the project named after the consultant.


## SET UP A BUDGET

The program budget should include all pertinent items listed in Figure 31 plus any others you might need. The biggest single cost will be the team. Other significant items will be temporary or part-time clerical help, additional computer hardware, purchased software packages, overtime, travel expense, consultant's services, and education. Don't overlook the last—just getting involved in a program doesn't make a person qualified any more than getting involved in sex makes someone a wise parent. The education budget should cover both outside courses for team members and in-house programs for all involved in or affected by the program. These should include foremen, storehouse supervisors, office managers and others who will use the system and who could have a significant effect on the benefits. Figure 32 is a matrix developed by Industrial Nucleonics project team showing in-house education for their people.

Be liberal with this budget. Compared to the benefits, it will be small. How else can you make an expenditure which will simultaneously reduce the capital investment and improve the return on it?

## TAKE CHARGE OF SYSTEMS DESIGN

Consider using available computer software packages. They can save you very substantial amounts of time, and help you avoid some serious errors. These two benefits alone will more than offset all the problems. Your data processing group will be quick to point out their disadvantages. Sure, they'll use a different programming lanuage. Of course, your people will have to spend some time becoming familiar with what's in the package. No, they won't produce exactly what the users think they would like to have. Yes, they'll cost you something out of pocket when you already have the staff and the computer to do the job yourself. But if you select a tested program, it can *do everything you really need done,* avoid some serious mistakes, and get the benefits showing up in the P and L statement months, maybe years, ahead of an in-house program. You don't leave the final selection of production equipment to the manufacturing engineers; why let the computer experts have the final word on the system machinery?

There are many software packages on the market. A few are excellent, some are good, some bad. I have no preferences—the real criterion is whether or not they work. The best way to find out what's in them and how well they work is to talk to a user. This may be tough on new packages, but my feeling is let someone else be the pioneer (that's the fellow out front with the arrows in his chest).

Some really unique packages are now available. These include not only the software programs, but also the computer and related hardware, and education and guidance in installing and operating

|  | Overview | Timetable | Cycle Counting Procedures | Unplanned Transactions | Material Ordering | Job Order Handling | Purchase Order Handling | Receiving Transactions | Inventory Transactions | Shortage Reports | Special Requests |
|---|---|---|---|---|---|---|---|---|---|---|---|
| **Production Planning** | | | | | | | | | | | |
| Supervisors | X | X | X | X | X | X | X | X | X | X | X |
| Material Analysts | X | X | X |  | X |  |  | X | X | X | X |
| Cycle Counters | X | X | X | X |  |  |  | X | X |  |  |
| Receivers | X | X | X |  | X |  |  | X |  |  |  |
| Stock Clerks | X | X | X | X |  |  | X |  | X | X | X |
| Shippers | X | X | X |  | X |  |  |  |  |  |  |
| Expediters | X | X |  | X |  | X | X | X |  | X |  |
| Dispatchers | X | X |  |  | X | X |  |  |  | X |  |
| **Other Depts.** | | | | | | | | | | | |
| Mfg. Managers | X | X | X |  | X | X | X |  | X | X | X |
| Q.C. Manager | X | X |  | X |  | X | X | X | X |  |  |
| Cost Account's | X | X | X | X |  | X | X | X | X | X | X |
| Purchase Buyers | X | X |  |  | X |  | X | X |  | X |  |

Figure 32 / Educational Program.

the system successfully. These will be particularly helpful to smaller companies and small divisions of large ones who do not have or do not want to get involved with big computers and systems, programming and data processing staffs.

They will also free many manufacturing managers from the frustrations of working through the large corporate systems staff. Plagued by poor knowledge of their capacity, little information on future loads, wasteful use of their resources, rigid formal procedures, and priorities favoring the "scorekeeping" activities, corporate systems staffs have been largely ineffective. Their cost has far exceeded the benefits. The real payback comes from improved manufacturing control, but corporate systems groups have impeded progress, not helped it. In my Keynote Address to the All-Wisconsin Conference of the American Production and Inventory Control Society in January, 1973, I predicted that "the days of central corporate systems groups are numbered." They cannot survive. *A system improvement must be designed, installed and operated under the user's control*—when he wants it, how he wants it, not when Big Brother gets around to doing it for him.

I'm not predicting the demise of large corporate computers. The "computer utility" is sound. Just like the electric and telephone utilities, it can provide power and services economically. How these are used must be determined by the "customer." Your electric company doesn't tell you what kind of toaster or how many lamps you need. You decide whether or not you can afford air conditioning. Why let your central systems group dictate—or even significantly influence—when and how you get better control of inventories and production?

## SET CHALLENGING GOALS

It's very hard to develop a tight timetable when even the major project details are not clearly specified. The tendency is always to throw in cushions of time to take care of unforseen difficulties. Also the experience of many other companies might lead you to believe that long, involved projects are normal. *But don't accept over-cautious timetables.* A payback of $500,000 per year means deferring over $40,000 in profits for every month the project is delayed. It's worth spending this much extra for each month cut from its schedule to develop the project. Avoiding long projects has an additional benefit. The team will be unable to waste much time playing around with frills and fancies. These can always be added later when the stripped-down system is producing the bulk of the benefits—if the cost of the extras can then be justified. You first need good dependable transportation; leave the stereo-tape player, cruise control, and rear seat cigarette lighter for installation later if you still think you want them.

## REVIEW PROGRESS REGULARLY

Set up a steering committee or a top management review board to review the progress of the team at regular intervals. As the details of the project firm up, request revised payback figures and program costs. Challenge the progress made and be intolerant of delays. These reviews have two functions. The obvious one is to keep the team on target and be sure they are moving steadily ahead. Equally important, however, is keeping top management aware of the benefits of maintaining high priority on such projects. They are often tempted by other problems or opportunities to go off on tangents, and this invariably diverts some of the effort needed to complete the program.

It's better to have two or more top management people on the steering committee. If there is only one, an organization change can bring a new manager on the scene who is unfamiliar with the project and its benefits. At best, it could take some time to familiarize him with the plan and its importance to the company. At worst, he could change priorities and drastically slow the project down.

Improved manufacturing control systems should be of as much interest to all top management as building a new plant or bringing out a new product line. When you develop new plants or products you go all out to stay on schedule, you don't interrupt the project in mid-stream or divert much effort into something else. Shouldn't manufacturing control systems, which can earn a higher return on investment, be treated the same way?

After embarking on a program to improve control, some companies cut off or reduce expenditures for the program because of a dropoff in sales. I think this is shortsighted, to say the least. Improved control of manufacturing operations is going to cost money sometime. Whether you call it an expense or an investment, it will have to be made sometime. Why not spend it now and get the payback sooner?

For one thing, it is even more important to have better control of operations when times are bad. Then too, people are not quite so busy drowning and have more time to save themselves.

Reduced inventories are always one of the major elements of payback. You get this benefit only when inventories are replenished at a slower rate than they are used up. The ideal time to implement a system and get it operating effectively is when business is slow. When the pickup comes, assembly rates can be increased to meet rising customer demand but component production rates can be held down until inventories have dropped to desired levels. This captures the savings at the earliest possible moment. It also eliminates other potential problems which can delay the payback. Try to convince a plant manager that foundry or machining operations should be cut back at a time when business is booming. It may be obvious that inventories cannot be reduced any other way, but manufacturing people see much more clearly the problems of changing capacity—twice!

## GET SOMETHING GOING

Implement first those system changes which generate the bulk of the payback. You will always get 80% of the results from 20% of the effort. Usually some manual, rough-cut methods of capacity planning and input/output control can be developed quickly to help bring work-in-process under control early in the program. This carries an added dividend. Work-in-process inventories going down offsets increases in finished goods inventories or components ready for final assembly. These increases usually result from the new system's ability to clear up shortages quickly. By cutting work-in-process you can get the improved customer service you want without any net increase in investment.

If you need material requirements planning, push for early completion of this program. Even if you can't get all of the bills of materials straightened out, get accurate bills for the vital few product lines. One company reduced inventories by more than $100,000 and improved customer service to over 98% in stock with a manual material requirements plan on only one product family.

Offsetting this "cherry picking" will be the need to get some programs going which appear to have no immediate payback. Typical of this is improved record accuracy. Unless you are really unique, you will have to get better control of storerooms, inventory transactions, and the other factors contributing to inventory record errors. And don't overlook open order files, bills of material and production reporting while you're straightening up the records.

Tangible savings from such programs are tremendous but intangible benefits may be equally important in the long range. How much is it worth to have confidence in decisions reached with sound, timely data? How much will a company's growth rate and competitive position be improved if lead times are short and it can react quickly to customer changes? What's the added payback from introducing a new product line in two months instead of six? What is the effect on share of the market of working off excess order backlogs (resulting from a sudden pickup in sales) in three months instead of fifteen? What's it worth to your company to get control of your operations six months before your competitors rather than six months after? What better investment can you make?

# 9

# Getting Maximum Control of Operations

## A SOUND SYSTEM IS A NECESSITY

Only with a sound body—a complete skeleton and a full set of muscles—can a man achieve full effectiveness; only with a sound system—all elements and a full set of smoothly working techniques—can a company achieve full control of its manufacturing operations. Of course, I'm speaking of the formal system. Every company has a system—or two or three or four—which they think they're using to run the business. These may even be effective in the hands of experienced people in stable times when nothing changes very dramatically. With hard work and some luck, the experienced people are able to keep production flowing, and meet most of management's goals.

Any significant changes, like adding some new products, sharply increasing demand, or a serious effort to reduce inventories, causes the delicate balance to be upset and the business goes out of control. In spite of more overtime and harder work by more people

149

scurrying around, things just don't happen as planned. Customer service gets worse, inventories go up and profits go in the other direction.

Then the finger-pointing starts. Each group is sure they are doing all they can, so it must be somebody else's fault—the forecasts are wrong, promises to customers by Sales are ridiculous, the shop never works to the schedule, the product design is faulty, Quality Control is super-critical, the customers are always changing their minds—as if these were something new.

The real tragedy is that there is no consensus on what the real problems are and what's needed to solve them. The absence of a capable, formal system makes it extremely difficult to identify the real problems so that management can get that concerted action so necessary to solve them. There's no substitute for a formal system capable of running the business, pointing the way so the organization can pull together and get itself out of trouble.

No formal system can function, however, unless it has a solid foundation of accurate records. People, not systems, make errors, and getting accurate records requires better management, not better systems. The route to accurate records is clear, although it is not easy to follow:

1/Determine how bad the records are by auditing regularly.

2/Estimate the costs of errors: out-of-pocket, tangible costs.

3/ Set goals and a timetable for improvement: realistic yet challenging.

4/Assign responsibility for correcting basic problems: specifically, with penalties and rewards.

5/Follow up, don't give up.

The critical records are well-known. Don't count on getting control until your people can depend on stock status data showing what's really on hand and on order, bills of material accurately describing what you're producing, routings showing correctly how you are making it, good time standards for how long it takes and good costs. These are the facts on which they are going to base control decisions. *Don't be satisfied until these basic data are in one file, updated by one system, and made accessible to all who need the information.*

When the formal system is faulty, it is necessary to prop it up with sub-systems. Then hot lists, staging areas and "little black books" run the operation instead of the formal system. Preventive maintenance and rebuilding programs are widely used for maintaining productive equipment. They are just as necessary for manufacturing control systems, which also get worn, go out of adjustment and break down. They too need some attention to keep them sound.

Sound systems, however, are not enough; good tools don't guarantee a good job. Most companies could do a lot better if they managed their present system properly. This is true even where it has some significant defects. How well information is used is much more important than where it comes from. Poor tools skillfully used can produce better results than fine tools poorly handled.

Managing with a system requires a different approach. Some managers never learn it. They refuse to accept limits on the freedom of choice they think they have—but they don't recognize the limited number of real alternatives available to them. The INSANE cycle I referred to in Chapter 1 is a classic illustration of this. Even management by objectives, as sound as it is, fails when it aims at each objective one-at-a-time instead of at balancing the three different and conflicting objectives. The real question is not one of getting inventory down, getting customer service up or improving profits—it's what *balance* you want among these three and how soon you want it.

## PLANNING FOR CONTROL

In Chapter 3 I discussed the widely prevalent idea that competition among managers is supposed to be good, and emphasized the need for common goals. Unfortunately, we see more competition between managers within a company than between them and their competitors in other companies. Marketing people seem to spend most of their time trying to outfox their own manufacturing people, rather than finding ways to outwit their competition. Engineers too often are working in their ivory tower developing new products or redesigning existing ones with little regard for standardization, inventory or manufacturing costs. Cost accountants scurry around trying to keep score on everything going on, but managers have to "engineer" the costs they need for making important decisions like changing capacity or reducing work-in-process. Data processing and computer people live in their own high-cost, specialized world full of promises for the future but with a history of poor performance (usually blamed on others), particularly in manufacturing control systems.

A basic business plan is required to get these groups working constructively together to meet common goals. The major goals are easy to identify. They are:

1/Profit levels needed to attract and hold capital for growth.

2/Capital needs for plant and equipment.

3/ The total customer demand anticipated and its breakdown among product lines.

4/The proper inventory levels considering the desired return on this investment.

5/Customer service levels needed to meet competition and get the desired share of market, considering also the inventory levels involved.

6/ The stability of employment balanced against the need to increase or decrease production output.

7/The need for innovation—new products, new processes and new systems—to keep the business alive and growing.

Management policies have to be established in each of these areas. The role of the system is to provide information on alternatives. In the hands of professional managers, a sound system can provide tradeoff data, enabling these policies to be based on rational analysis rather than on the selection of "magic" numbers. Examples were given in Chapter 5 illustrating this.

## CONFUSING THE SYSTEM

Translating the overall business plan into more detailed material and production plans is accomplished through the master production schedule. This specifies what is to be made, how many and when, and drives the material ordering system. Managers find great difficulty in resisting the temptation of telling the schedule lies. At some time, every company with an effective system has fallen victim to its own optimism in setting production schedules. Instead of telling the truth, *what they were capable* of making, they told the system *what they would like to be able to make.* Their systems then lied back to them (at electronic speeds) about what materials they needed and when they were required.

Managers also find it difficult to face up to the reality of controlling lead times. They try to "make life easier" by using inflated lead times. This simply extends the minimum horizon for the master production schedule, making it less likely they will know what they will really need that much further in the future. This causes the system to generate false priorities and more work for their own plant and for their vendors, making the job of getting the right items on time that much more difficult.

Another temptation that management seems to find irresistible is adding safety stocks at most inventory levels. If this only increased the inventory investment it would be bad enough, but even more harmful, this really destroys the validity of work priorities and the credibility of the people using these priorities that the system is telling them the truth about what they should make.

A system is really a mirror reflecting in different terms what you have told it. Valid output requires valid input. After a sound master production schedule has been generated, the system's role is to translate this into specific plans to get the capacity and follow the priorities to make the products on the master schedule. It's really just that simple. The system should tell you whether or not you are making enough, and if you are working on the right things today. Of course, this is much more easily said then done. The bulk of this book has been concerned with techniques for priority and capacity planning and control. I have emphasized that both are necessary, and that they interreact directly and inevitably. The system shows you either what resources you need to make what you'd like to, or what you can make with the resources you have. It reflects what you tell it. It doesn't add or subtract magically to give the answers you'd like. *Either get the capacity, or tell yourself and your customers the truth about what you will produce.*

Systems can generate far more information than managers can digest. Properly designed, they can produce exception reports or action notices highlighting only those measures that are outside the acceptable tolerance limits. Management can help itself by setting realistic tolerance limits so that it sees fewer deviations. In the end, though, it is the role of the manager to sort the vital few facts from the trival many, and to determine which will get their attention today.

## IMPORTANCE OF DISCIPLINE

The staff of life of systems operation is discipline, not in the sense of punishment, but in the sense of orderly, dependable execution of systems activities. The most significant difference between manufacturing control here in the United States and abroad is in discipline. The foreign operations I've seen can depend on the system being operated as it was intended. We seem to love to tinker, set up handy-dandy sub-systems, and patch it up rather than really fix the formal system. On the other hand, Europeans work hard getting the most from what they have. As a result, Europeans live in system "homes" that are uncomfortable, inflexible, and old but well-maintained, while we in this country have patchwork, ramshackle structures we're trying to prop up faster than they fall down.

Unfortunately, discipline seems to be the first victim of crisis. Fire-fighting always has higher priority than fire-prevention, and manufacturing people seem to find it much more to their liking. There are ways, however, to insure regular performance of the routines necessary to keep the system functioning and up-to-date. One such tool is a checklist of regular chores which lists those things which must be done periodically. Such activities as "up-date master production schedule," "review and release new orders," "followup past-due orders," "review excess orders to be cancelled" and other routine things could be listed and checked regularly to insure they are done when needed to keep the system in good order. Making up such a checklist  generates double benefits. Thinking through the system to identify the important routines develops a much better understanding of its fundamental requirements. Having supervisors review such checklists regularly with their people not only insures that the routine work is done, but also brings to light the problems involved in doing it. The checklist approach results in more time spent planning to stay out of trouble, and less reacting to get out of it.

## ORGANIZING FOR RESULTS

In Chapter 3 I discussed the relation between the form of organization and effective results from better control. I concluded that setting the proper primary goals, getting common effort toward mutual objectives, and having good information to work with was more necessary for success than the form of organization used. In fact, the organization structure is almost irrelevant.

To fix responsibility for falldowns, it is necessary to define the roles of organizational groups involved in manufacturing control activities from the standpoint of planning versus execution. When the specific goals in three primary objectives—customer service, operating costs and capital investment—are not met, the answer to the question, "Who's responsible?" clearly depends on whether the planning or the execution was at fault. It's just as common to see materials managers fired for product shortages as it is to see sporting team managers fired for a lack of victories. It is interesting to note how many of each turn up as heroes on another team. The real problem must be identified or it's not likely to be solved.

It's surprising how few companies have attempted to define the primary roles of major organizational groups. Here are my definitions for some of the more important ones:

1/ Marketing: to develop information on the potential market for the company's products, including as much detailed information as possible on demand from specific customers and territories, and the effect on this demand of various product models and prices. Actual selling of the product should be secondary to providing information to the rest of the organization on demand for products.

2/ Manufacturing Control: to develop realistic plans for producing the necessary output, measure progress against these plans on a timely and accurate basis, and report significant deviations for corrective action, all this consistent with meeting the three primary objectives.

3/ Manufacturing: to provide the necessary manpower and facilities or outside sources to produce the required total of product in accordance with schedules giving detailed priorities.

4/ Purchasing: to locate better sources for purchased materials and services, and negotiate favorable prices, terms and conditions. Clerical work, expediting, and other activities should be transferred elsewhere unless they contribute to better prices from better sources.

5/ Engineering: to design a product acceptable to the market which can be manufactured at a profit, and to provide detailed information describing it to the rest of the organization in bills of material, specifications, drawings, etc. Innovation, economy, and information should carry equal weight in engineering activities.

Such definitions are needed to minimize the internal conflicts so common in business. The manufacturing control system will tie all these organizational groups together in their planning activities. When the system is clearly understood, thinking about how each departmental group functions with it will help develop clearer definitions of their responsibilities. Devote your time first to these definitions. Then you'll need less time fitting people into organizational forms. People must see their goals clearly and recognize who must work with them to achieve these goals. When they are measured on how well they get there *together*, they will spend very little time worrying about who works for whom under the formal organizational chart. With good definitions and clear understanding, any organizational form works; without these, any organization is a hollow shell about as useful as a watch case without the works.

## THE IMPORTANCE OF PROFESSIONALISM

There is no substitute for qualified, professional people to run the manufacturing control system. Developments in this field have come fast in the last five years, and these developments have included some new techniques. The bulk of the change, however, has been in refinements of older techniques and in their successful application as part of the complete system. At the present time there are too few fully qualified professionals around. If you need one, don't count on hiring him away from someone else. You'll have to develop your own. Contrary to much popular opinion, however, professional manufacturing control can be taught. It is not necessary to learn just by experience. In fact, experience is probably the poorest teacher, as I demonstrated with my "Common Sense Rules" in Chapter 1. Keeping in mind the principles discussed in this book, it's easy to see clearly why each of the six rules is doomed to failure.

1/ "If a little expediting is good, a little more will be better." This attempts to substitute more chasing around for a formal system capable of doing the job. This is like getting a better wheelchair rather than strengthening weak leg muscles to make them capable of walking unaided.

2/ "To get more production from a factory you must put more orders into it." This ignores the input/output principle of never releasing more work than was turned out in the preceding period. Additional capacity requires more people working more hours, or higher productivity rates, not more open orders in the plant.

3/ "To get important jobs completed on time, get them started as soon as possible." This also ignores the input/output principle, indicating failure to comprehend that lead time is controlled only by having adequate capacity and a good priority system, not simply by giving everybody more time.

4/ "If the planned lead time isn't long enough, increase it." This also overlooks the interaction between capacity and priority. The cure is working off excess material-in-process to bring planned and actual average lead times back into balance.

5/ "Lay out (stage) sets of parts further in advance of assembly if you need more time to expedite shortages." This is another attempt to build a stronger sub-system. But then some people prefer wearing a tighter girdle to reducing and exercising to get their figures under control.

6/ "If you're short several items made on one machine, get out of trouble quicker by cutting all the lot sizes." This attacks the wrong problem. The symptoms indicate a priority problem, but the disease is lack of capacity. These are frequently confused by the amateur. He gets deeper in trouble by cutting order quantities, thus making more changeovers necessary and reducing capacity even more.

The qualified professional does not make these mistakes or allow his boss to pressure him into making them, because he understands what is symptom and what is disease.

Education in the field of manufacturing control is becoming more widely available. The technical society of practitioners, the American Production and Inventory Control Society (APICS), has been growing steadily in its first two decades of existence. Almost 150 Chapters located throughout the world hold monthly meetings and annual seminars, which are forums for the latest thinking and experience in the field. Regional groups hold seminars, and the Society conducts annually an International Conference. These all provide the best available sources for the practitioner seeking information and contacts to help him in his job. The publications of the Society are the best sources of technical information available. These include the Production and Inventory Control Handbook, the Quarterly Journal, Annual Conference Proceedings, Special Reports, Training Aids, a Dictionary, and a Bibliography of articles in a wide spectrum of trade journals. Beginning in 1973, the Society is conducting certification examinations which provide every practitioner with an opportunity to evaluate how well he understands the body of knowledge in his field. These will cover five modules: Forecasting, Inventory Planning, Material Requirements Planning, Capacity Planning and Control, and Shop Floor Control.

Universities, colleges, consulting firms, trade associations, computer manufacturers and various individuals conduct courses on manufacturing control. This is one field where education need not be only a long-term investment. I have seen many men put to work immediately some technique they learned from these courses that paid off the learning expenses in a few months. Educational programs can easily be self-supporting if they are properly designed and if the right objective is set. *The goal should be to get something useful working today, to* do something different; *don't just get better educated.*

Educational programs need not be dependent solely upon outside leadership. In-house sessions developed and led by a company's own personnel have been among the most effective programs I have seen. Specific topics such as "Preparing the Master Schedule," "Controlling Capacity in Service Departments," and "Cycle Counting Problems" can be assigned several weeks in advance to one individual who prepares to lead the discussion, not to make a formal presentation. The topics are not discussed theoretically but as they relate to the company's operations. The discussion leader learns more than anyone else, but many ideas for improvements are generated among the group. This is almost literally lifting yourself up by your boot straps. George Bernard Shaw said, "Those who can, do. Those who cannot, teach." I've found that those who do can teach if motivated and assisted. The costs of such programs are negligible; the benefits invaluable.

Other low cost educational tools are also available but rarely used. Trade journals frequently have articles on pertinent subjects. These can be circulated with requests to specific individuals to comment on possible applications. Meetings of other technical groups like the Purchasing Agents, Industrial Engineers, and Data Processing frequently discuss manufacturing control topics. Why is it so few companies send people to such meetings seeking answers to specific questions? An individual attending such a session should be planning the report he will make when he gets back. I am constantly amazed at how few take notes at technical society meetings. Obviously, most of those attending have no real objective nor any requirement to report on what they heard and saw when they return to their companies. This is rank waste of a fine educational opportunity.

Systems documentation can also be used for educational purposes. The user documentation can be the basis for a group discussion to acquaint all individuals with how the system is supposed to work. The discussion can also serve as a check of how accurate and up-to-date the documentation really is in describing what is actually going on.

## MEETINGS

Meetings are a vital part of managing with a system, but few are well-run. A clear consensus of managers indicates that practically all meetings are wasteful and largely ineffective. While meetings are inevitable, they don't have to be interminable. Meetings are too long and unproductive because:

1/ There are too many people present.

2/ They represent too many different levels of management.

3/ The definitions of basic responsibility and authority of those attending are not clear.

4/ There is inadequate preparation, no agenda, and facts are incomplete.

5/ The focus is on symptoms rather than diseases.

Meetings have only one purpose and that is face-to-face communications. Individuals must make decisions. Their decision-making ability can be helped rather than hindered only if meetings are structured properly. The basic objective is to get decisions made promptly at the lowest possible level of the organization. A series of meetings should be organized, therefore, starting at the foremen and plant manufacturing control people's level, to handle all problems within the scope of their authority. These would include overtime, detailed schedule changes, quality problems, and the like. Information on tooling difficulties, machine breakdown, engineering problems, and other matters outside their scope should be brought to middle management levels for decision and action. Problems involving customer relations, investment in plant and equipment, and significant deviations from expense budgets or inventory investment goals should occupy the attention of top management. A hierarchy of meetings is needed. The series of meetings should follow each other promptly to insure effective use of management time and prompt solution to problems.

The A-B-C principle applies also to problems. At each level of management there are a vital few problems whose solution today would bring the greatest benefit to your operation. Almost invariably these "A" problems cannot be solved in one department or by one manager. Solving the top problems requires a concentration of effort on each problem by the whole team. This will never be achieved by flooding managers with information of all kinds, hoping they'll all extract the same conclusions. Attention must be focused on specific problems. Concentration of effort can only be achieved by the correct use of regularly scheduled meetings, with an agenda prepared properly for these meetings.

The manufacturing control system is an *information system* for planning and control. The Production Control, or Materials Manager or whatever you call the individual in charge of this control system, should have the responsibility for developing an agenda for the regular meetings. The general format illustrated in Figure 33 has been found extremely useful for such an agenda. It should be distributed well in advance of the meeting to the individuals who will attend. The first section is a summary of performance versus major goals at the level of organization involved. Obviously, these should include detailed information on customer service, forecast demand, inventory levels, capacity measures and schedules. The vital part of the agenda, however, is the listing of major problems. Every attempt should be made to identify diseases,

Distribution
Team

Perfomance vs. Goal Summaries
Customer Service
Demand Forecast
Total Inventory
Plant Capacity
Record Accuracy

Summary of Major Problems (5-8 only)
Problem
Recommendation
Follow-up

**Figure 33** / Agenda—Manufacturing Control Meetings.

not symptoms. A very workable approach is to ask a series of at least four "why" questions. Customer service is not up to goal. Why? These additional items just appeared on the out-of-stock list. Why? Total output of this department has been inadequate to meet the customer demand. Why? Quality problems are interrupting production. Why? Usually after four "why's" you are very close to the real problem.

The total number of problems listed on the agenda should not exceed eight. No management team can attack and hope to solve more than a few problems each week. A recommended solution should accompany each problem, just to get the discussion going. Foremen, quality control people, engineers and others specifically involved should be contacted in advance to get agreement on the facts. This eliminates surprises during the meeting, the defensive reaction of people put on-the-spot unexpectedly. It also provides an opportunity for responsible individuals to decide on or begin taking some corrective action *before* the meeting so that they appear as "heroes" instead of "goats." Since decisions are not actions, problems should stay on the list until they have been corrected. No single technique I know of can pull a management team into better working operation than this one. There is no finer tool for focusing attention, getting concentrated effort, and solving problems promptly.

The information control system used properly can bring out true facts about the operation. These must then be analyzed and a course of action decided upon. Alternative actions are not always readily apparent, are usually fewer than managers would like, and are almost always unpleasant. "Making the least worst choice" seems to be the name of the manufacturing control game. When capacity is inadequate to meet total demand, for example, it is obvious that some customers are going to be unhappy, or that capacity must be increased. These are the only real alternatives, however unpleasant they may be. If the decision is made to keep capacity at present levels, then the "alternatives" become selecting which customers will be made unhappy. This is even more unpleasant. It is the role of manufacturing control to point out such real alternatives. In every such decision some manager will find valid reasons for criticizing it. Manufacturing control is not a popularity contest, however, and as President Harry Truman once said, "If you don't like the heat, get out of the kitchen."

## THE PAYBACK

Interest in developing better systems has been growing rapidly in manufacturing companies. This is undoubtedly generated in part by success stories which have been published, documenting the benefits some companies have achieved. It also results from manufacturing control "coming of age" and receiving the recognition it deserves. *Professional manufacturing control managers are being heard.* They know how to design a formal system capable of providing the information needed to manage a business better. This progress is not limited to a few specific companies but has been achieved in both small and large companies, old established firms and new ones just getting started and, believe it or not, companies without computers as well as those with lots of computer horsepower.

More and more companies are appearing in the literature as successful at improving manufacturing control. Twin Disc, Black & Decker, Markem Corporation, New Britain Hand Tools, Perkin-

Elmer, Dodge Manufacturing, The Franklin Mint, Hesston Corporation, McGill Manufacturing, Ivan Sorvall, Bendix, and a few others have made significant gains. Some specific data has already been given in Chapter 1 indicating the magnitude of these gains. Much more could be added. Here are some other interesting statistics:

1/ Eight months after their control system was implemented, this company reduced $2-1/2 million in inventories by more than $500,000, and simultaneously improved customer service from 65% to 84%.

2/ While reducing safety stock inventories by 75%, this company cut open order backlogs 65% and shortened manufacturing lead times from 10 weeks to 5 weeks.

3/ While holding total inventories constant, this company maintained customer service at the 94/95% level in spite of a sales increase of 25% in one year. This was far in excess of the growth rate in their total market, indicating that they were taking substantial business from competitors.

4/ After the sharpest sales growth in this company's history, unshipped order backlogs returned to previous levels in three months. This used to take fifteen months before they improved their manufacturing control system.

5/ Because they have no computer, this company rented another company's computer system, adapted it to their operations in only four months, and improved ontime deliveries very significantly, holding total inventories at the same level while sales increased 40%. After a major surge in sales, they restored back-orders to previous levels in only four months.

6/ In the face of a 28% growth in sales over two years, this company improved turnover of total inventories from 4.9 to 7.0, saved $100,000 in premium transportation charges, and more than doubled return on investment.

What else can be done in manufacturing industries to achieve anything like these results? Technological developments in new processes and new materials will continue to produce improved products and help reduce costs. New products and marketing techniques will open up new sources of income. The potential benefits of all of these, however, are relatively small compared to those from improved control of operations. In addition, their gains may easily be nullified unless control is achieved. What are the alternatives to effective control? Continual tinkering with informal systems, patching and propping up the formal one, recurring crises, dissatisfied customers, too much inventory and too little profits. Effective control brings significant rewards, but don't view it simply as a boon. It may soon be a necessity. With it, your company is sure to make some gains. Without it your company may lose more than the benefits—it may not survive.

# Bibliography

1. Harris, F. W., "Operations and Cost," Factory Management Series, A. W. Shaw Company, Chicago, Illinois, 1915.

2. Harty, J.D., G. W. Plossl and O. W. Wight, "Management of Lot-Size Inventories," APICS Special Report No. 1, 1963.

3. Welch, W. Evert, *Tested Scientific Inventory Control,* Management Publishing Company, Greenwich, Connecticut, 1956.

4. VanDeMark, R. L., "Hidden Controls for Inventory," Proceedings 1962 APICS Conference.

5. Brown, R. G., "Aggregate Inventory Management," Proceedings 1962 APICS Conference.

6. Plossl,G.W., and O. W. Wight, *Production and Inventory Control: Principles and Techniques,* Prentice-Hall, Englewood Cliffs, New Jersey, 1967.

7. Wight, O. W., "Input/Output Control - A Real Handle on Lead Time," Production and Inventory Management, *Journal of APICS,* Third Quarter, 1970.

8. Wilson, R. H., "A Scientific Routine for Stock Control," *Harvard Business Review*, Vol. 13, No. 1, 1934.

9. *Newsletter* #5, "Vendor Delivery Problems? Are You Victim or Culprit?" G. W. Plossl & Company, Inc., Decatur, Georgia, 1970.

10. Magee, J.F., and D. M. Boodman, *Production Planning and Inventory Control,* McGraw-Hill, New York, 1967.

11. Greene, J. D., *Production and Inventory Control Handbook,* McGraw-Hill, New York, 1970.

12. Putnam, A. O., E. R. Barlow and G. N. Stilian, *Unified Operations Management,* McGraw-Hill, New York, 1963.

13. Theisen, E. C., *How to Obsolete Inventory Obsolescence,* G. W. Plossl and Company, Inc., Decatur, Georgia, 1972.

14. Plossl, G. W., *MRP and Inventory Record Accuracy,* G. W. Plossl and Company, Inc., Decatur, Georgia, 1972.

15. Orlicky, J., *The Successful Computer System,* McGraw-Hill, New York, 1969.

16. Plossl, G. W. and O. W. Wight, *Material Requirements Planning by Computer,* APICS Special Report, 1971.

17. Townsend, R., *Up The Organization,* Alfred Knopf, New York, 1970.

18. Wight, O. W., *The Executives New Computer,* Reston Publishing Company, Reston, Virginia, 1972.

19. Plossl, G. W., *"How Much Inventory is Enough?"*, Production and Inventory Management (APICS Journal), Second Quarter, 1971.

20. Garwood, R. D., *Delivery as Promised,* Production and Inventory Management (APICS Journal), Third Quarter, 1971.

21. Gingrave, M. J. and G. W. Cuff, *"TIME Makes The Cost Difference,"* Production, August, 1971.

22. Quinby, E., *"Advanced Requirements Planning System Cuts Inventory Costs, Improves Work Flow,"* Production and Inventory Management (APICS Journal), Third Quarter, 1970.

23. Plossl, G. W., *"Managing The Production Control System,"* Proceedings, 1969 APICS Conference.

24. Janesky, A. J., *"Computerized Control in A Small Company,"* Proceedings 1969 APICS Conference.

25. Plossl, G. W., *"Getting The Most from Forecasts,"* Production and Inventory Management (APICS Journal), First Quarter 1973.

26. Newsletter #11, *"What Better Inventment Can You Make?"* G. W. Plossl & Co., Inc., Decatur, Georgia, 1971.

# Index